Lecture Notes in Earth Sciences 53

Editors:
S. Bhattacharji, Brooklyn
G. M. Friedman, Brooklyn and Troy
H. J. Neugebauer, Bonn
A. Seilacher, Tuebingen and Yale

Lecture Notes in Earth Sciences

55

Editors:
S. Bhattacharji, Brooklyn
G. M. Friedman, Brooklyn and Troy
H. J. Neugebauer, Bonn
A. Seilacher, Tuebingen and Yale

Frank Scherbaum

Basic Concepts in Digital Signal Processing for Seismologists

Springer-Verlag
Berlin Heidelberg GmbH

Author

Prof. Dr. Frank Scherbaum
Institut für Allgemeine und Angewandte Geophysik
Universität München
Theresienstr. 41, D-80333 München, Germany

"For all Lecture Notes in Earth Sciences published till now please see final pages of the book"

ISBN 978-3-540-57973-1   ISBN 978-3-540-48403-5 (eBook)
DOI 10.1007/978-3-540-48403-5

CIP data applied for

© Springer-Verlag Berlin Heidelberg 1994
Originally published by Springer-Verlag Berlin Heidelberg New York in 1994

Typesetting: Camera ready by author
SPIN: 10129898   32/3140-543210 - Printed on acid-free paper

For Knut †

For Kим

Foreword

Seismological data analysis is currently undergoing a major change. Digital data acquisition has become the standard method for recording seismic data. Strong national and international programs have been launched with the goal to increase the number of digital seismic stations worldwide. During the upcoming International Seismological Observing Period (ISOP), one of the major goals will be the enhancement of seismological observatory practice worldwide (Dornboos et al., 1990). In this context, on-site analysis of digital seismograms will play a central role. It is now possible for seismologists with as little equipment as a PC and a modem to gain access to high quality digital seismic data and to perform high quality analysis.

Dealing with digital data, however, doesn't make the analysis easier than the analog era. On the contrary, even basic tasks such as reading of onset times or amplitudes from seismic records may require quite a bit of programming effort, not to speak of tasks such as filtering, spectral analysis, polarization analysis, etc. In addition, the application of digital signal processing techniques is not always straightforward. Even with all the software nowadays available for analysis of digital seismograms, a basic understanding of continuous and discrete system theory is essential for the correct application of digital analysis tools. Focusing on a model of modern seismic recording systems as a sequence of analog and continuous, linear, time invariant (LTI) systems, this course introduces the basic concepts of digital signal processing in a very pragmatic way.

In order to create the examples and most of the problems treated within this text, I have been using PITSA, a program written by myself and Jim Johnson (Scherbaum and Johnson, 1993). The program is available as Vol. 5 of the IASPEI Software Library through the Seismological Society of America, 201 Plaza Professional Building, El Cerrito, CA 94530, USA (Phone: 510-525-5474: Fax: 510-525-7204). PITSA was primarily designed for research purposes and the application in day-to-day station-based seismological analysis. With the analysis tools becoming theoretically more and more demanding, however, we started out to add some educational features, mainly to test the behaviour and to teach ourselves the proper application of certain algorithms. This led to the first version of a *Short Course on the First*

Principles of Digital Signal Processing for Seismologists which started out as an undergraduate course at the University of Munich in 1991 (Scherbaum, 1993).

Since 1991, I have taught this course to different audiences in different environments. In March of 1992, it was given as a three-day block course at the Geophysical Institute of UNAM, Mexico City, and in 1993, part of the material was presented as a two-day Signal Processing Workshop in Baltimore, organized by the Incorporated Research Institutions in Seismology (IRIS). I have benefited enormously from the feedback and the enthusiasm of the participants of these courses. As a consequence, the original lecture notes have been completely revised and grown (especially on the theoretical side) to what has become this text. During all the revisions, I have tried to keep the concept of a *hands-on-let's-check-it-out* approach. Most importantly, solutions to the problems presented in the course have been added to the lecture notes.

Although most of the problems and examples in this course have been produced and are discussed using PITSA, access to this program is not of absolute necessity to follow the material presented. Readers with access to programs such as Math-CAD[1], Mathematica[2], or Maple[3] will find that they can reproduce all of the essential examples with only moderate programming efforts. For readers who completely lack access to any of these programs, PITSA screendumps for the most important steps of the solutions are given in "Appendix A: Solution to Problems" starting on page 111.

I am indebted to the students at the University of Munich who involuntarily had to test PITSA in a teaching environment. I have benefited from numerous discussions on signal processing with Axel Plesinger, Miroslav Zmeskal, Jim Johnson and Joachim (Jo) Wassermann (who also prepared the original manuscript for Frame-Maker[4]). Wang-Ping Chen deserves special thanks for his critical reading of the manuscript.

München, August 1993   Frank Scherbaum

---

1. MathCAD is a registered trademark of MathSoft, Inc.
2. Mathematica is a registered trademark of Wolfram Research, Inc.
3. Maple is a registered trademark of Waterloo Maple Software.
4. FrameMaker is a registered trademark of Frame Technology.

Contents

1 Introduction

By recording seismic ground motion, seismologists are trying to obtain information about physical processes within the earth. A central target of attention has historically being the earthquake source. However, ground motion recorded at a seismic station on the earth's surface differs considerably from seismic signals originated at the earthquake source (Fig. 1.1).

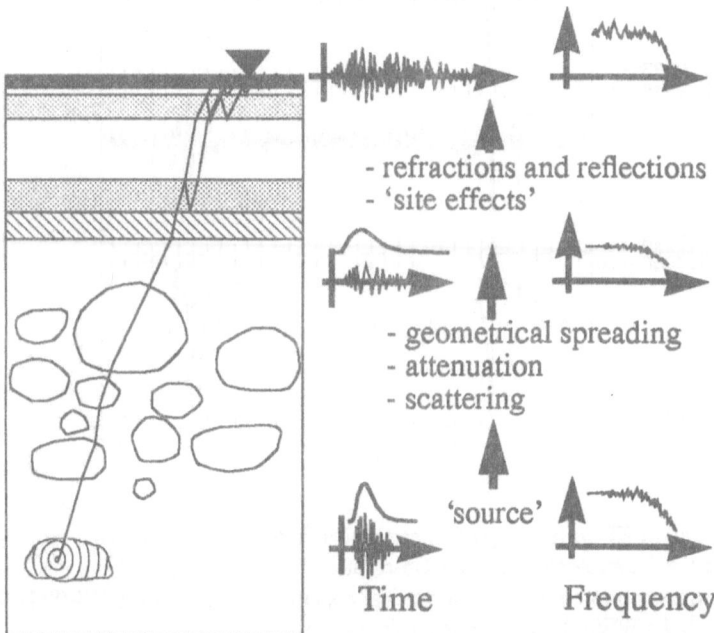

Fig. 1.1 Signal distortion during wave propagation from the earthquake source to the surface.

In Fig. 1.1, some of the conceivable changes in waveform along the propagation path are sketched schematically. Attenuation causes frequency dependent reduction of the amplitudes and phase shifts. Scattering will produce complicated superpositions of wavelets with different paths, and reverberation in shallow sedimentary layers will cause frequency dependent amplification. Finally, the recording system and the sampling process produce additional signal distortions.

Fig. 1.2 illustrates the influence of the recording system on the observed waveform by showing the same seismic signal as it would be recorded on three different instruments. The first visible wave group is PKP, a signal that has traversed the earth's core while the second one is pPKP, its surface reflection. Each instrument distorts the incoming signal in a different way, emphasizing different frequency components of the signal.

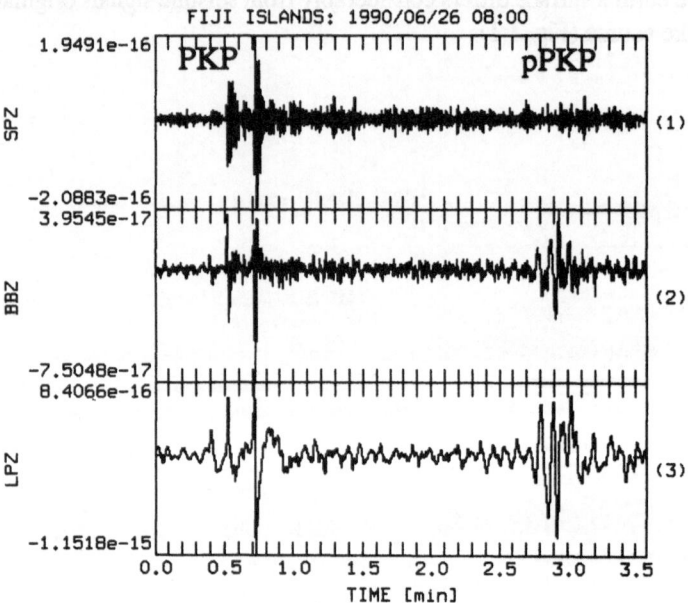

Fig. 1.2 PKP and pPKP wave group of an earthquake in the Fiji islands region recorded at station C1 of the Seismological Central Observatory Graefenberg (GRF) ata distance of 151°. Shown from top to bottom are the vertical component records for a: WWSSN SP, a KIRNOS BB, and a WWSSN LP instrument simulation.

Before we can even begin to think about the interpretation of recorded seismic signals in terms of properties of the source and/or the earth, we have to understand and possibly correct for the effects caused by the recording process. In the following chapters we will see how these effects can be understood and modelled in a quantitative way.

In addition to the effects of the seismometers, we have to understand the limitations of the data imposed by the sampling process. For maximum resolution, current state-of-the-art recording systems often make use of oversampling/decimation techniques. For reasons which will be discussed later, digital low-pass filters with two-sided impulse responses similar to the top trace in Fig. 1.3 are commonly used in this context. As a consequence, the onset times for very impulsive seismic signals may be obscured by precursors and become hard, if not impossible, to determine, especially forautomatic phase pickers.

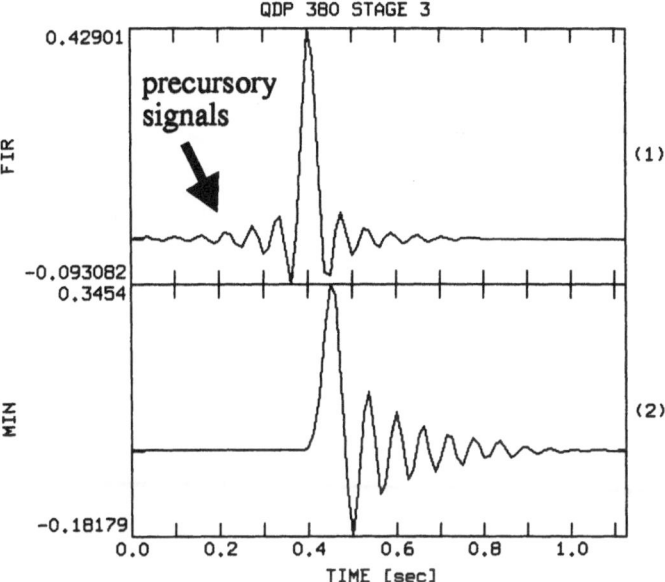

Fig. 1.3 Impulse response of stage 3 of the two-sided decimation filter incorporated in the Quanterra QDP 380 system (top trace). The bottom trace shows a filter response with an identical amplitude but different phase response.

In chapter 8 *The Digital Anti-Alias Filter*, we will see how to treat this problem theoretically as well as practically. We will learn how to remove the acausal (left-sided) portion of such a filter response from seismic signals for the determination of onset times. In other words, we will see how to change the two sided filter response shown in the top trace of Fig. 1.3 into a left-sided equivalent, as shown in the bottom trace of Fig. 1.3.

In the final introductory example, we are going to look at a typical sequence of signal processing steps to determine seismic source parameters from local earthquake records (Fig. 1.4).

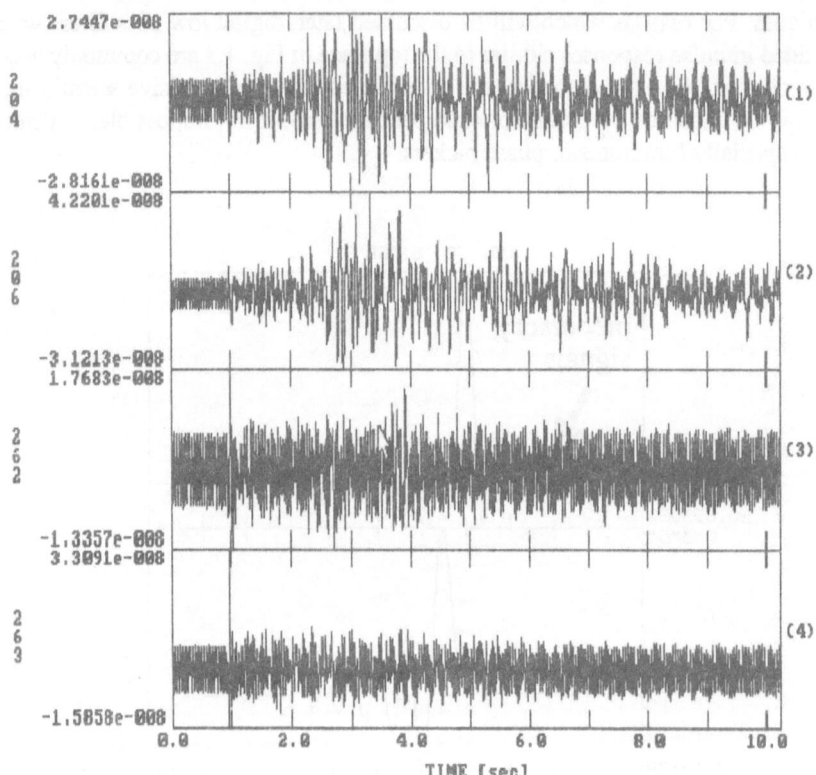

Fig. 1.4 Vertical component, velocity records of stations 204, 206, 262, and 263, respectively of event 4 of the Chalfant Valley aftershock sequence (cf. Luzitano, 1988). Notice the monochromatic noise that has been artificially superimposed on the data to simulate the effects of crosstalk.

Fig. 1.4 shows four short-period, vertical component records for an aftershock of the 1986 Chalfant Valley earthquake recorded at temporary stations of the USGS. Monochromatic 60 Hz signals have been artificially superimposed on the original records in order to simulate the potential effect of crosstalk. After the correction for the preamplifier gains, a typical first step of digital signal processing would be the removal of this and other kinds of unwanted 'noise' from the data. The techniques covered in this course will enable us to design and apply simple filters for this purpose. After noise removal, the records from stations 204, and 262 look like in Fig. 1.5

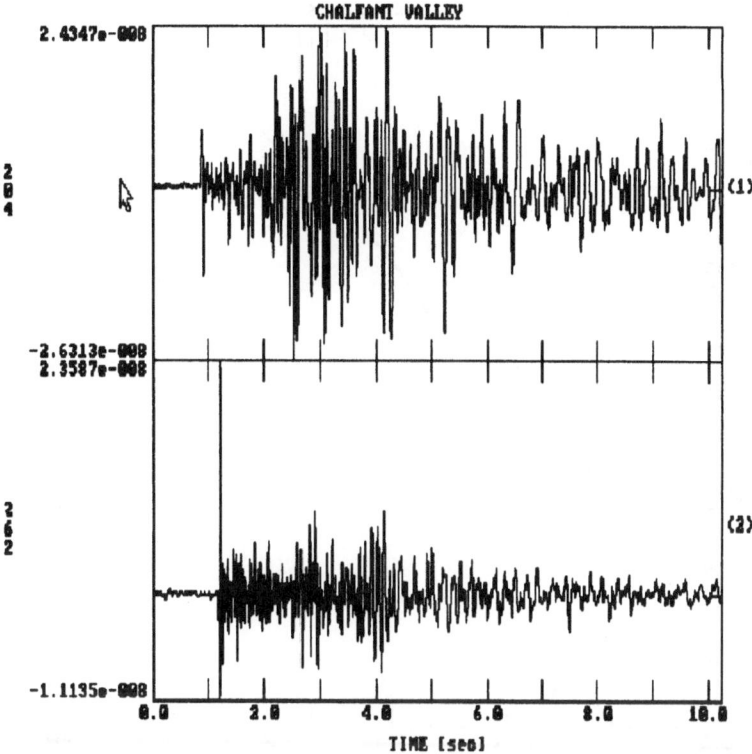

Fig. 1.5 Velocity records of stations 204, and 262, respectively, of event 4 of the Chalfant Valley aftershock sequence after noise removal.

What we ideally would like to obtain by applying signal processing techniques to the recorded data is either an approximation to the true ground motion or at least a simulation of records of some standardized instruments. In technical terms, the first problem is called *signal restitution*, while the second one is known as *instrument simulation* problem. For both of these tasks, the behaviour of the recording system must be described in a quantitative way. As two of the most powerful tools in this context, we will become acquainted with the concepts of the *transfer function* and the *frequency response function*. The modulus of the latter provides a very illustrative way to visualize the frequency dependent amplification and damping of seismic signals by a specific instrument. For the recording system used for the data shown above, the theoretical frequency response function (modulus) is shown in Fig. 1.6

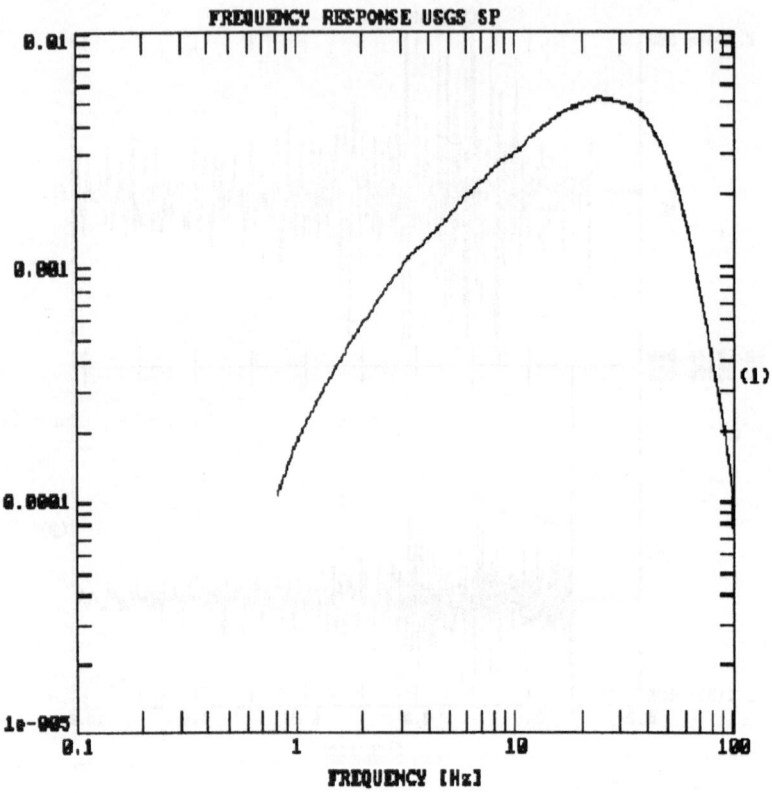

Fig. 1.6 The theoretical frequency response function (modulus) for the USGS short- period refraction system (normalized response to ground displacement). After Luzitano (1988).

After we learn how to describe the effects of seismic recording systems, we will also learn how to describe their interactions with the ground motion. This will lead us to the topic of *convolution* as one of the essential key topics of system theory. In the context of restitution and/or simulation, however, we are even more interested in the inverse process, which is called *deconvolution*. Again, there are many different deconvolution techniques with different advantages and drawbacks. Here, we touch upon the *spectral division* technique to illustrate some of the basic effects. If we apply this method to the data in Fig. 1.5, we will obtain the records shown in Fig. 1.7.

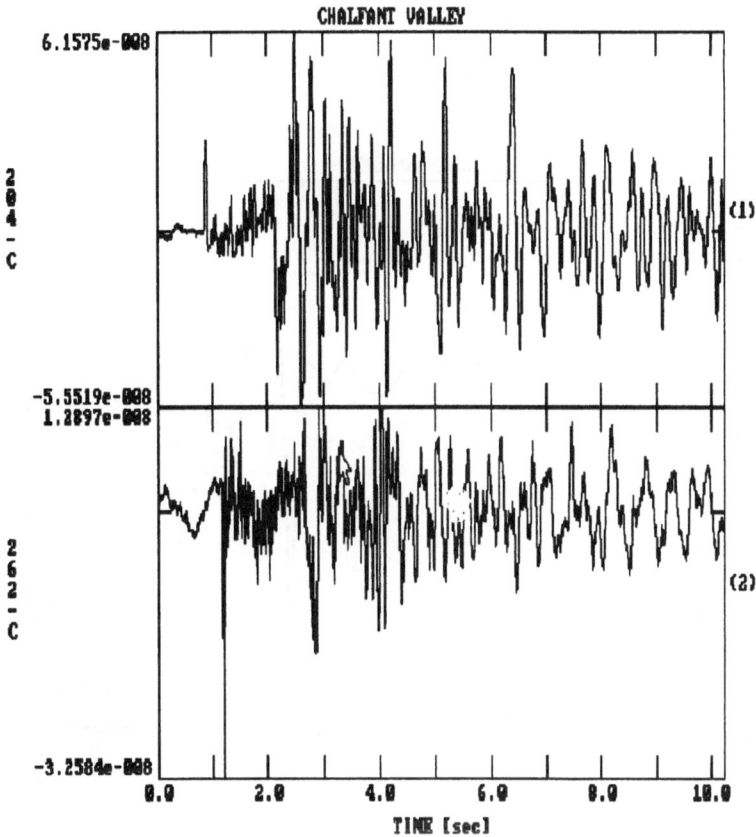

Fig. 1.7 Instrument corrected displacement records of the signals shown in Fig. 1.5.

Once the signals are corrected for the instrument response, they can be used for seismic source parameter inversion. What we conventionally try to determine from this kind of records are estimates of the seismic moment M_o, which is proportional to the product of source size (radius) and source dislocation, the stress drop, as well as of the attenuation properties along the ray path. In this context, it is quite common to do the interpretation in the *spectral domain*, by calculating the Fourier transform of the data. If we do this for the P wave portion of the top trace in Fig. 1.7, we obtain the *spectrum* shown in Fig. 1.8. Its shape is more or less typical for a displacement spectrum, with an essentially flat spectrum plateau (here roughly between 1 and 10 Hz), and a high frequency region in which the spectral amplitudes decay rapidly with increasing frequency (here above 10 - 20 Hz). The transition region is commonly described by the so-called *corner frequency*, f_c, defined as the frequency at which the spectral amplitude of the source signal is 1/2 of the plateau value. In a double logarithmic plot, the corner frequency appears roughly at the intersection of two asymptotes: one for the plateau and the other for the high-frequency region of the spectrum.

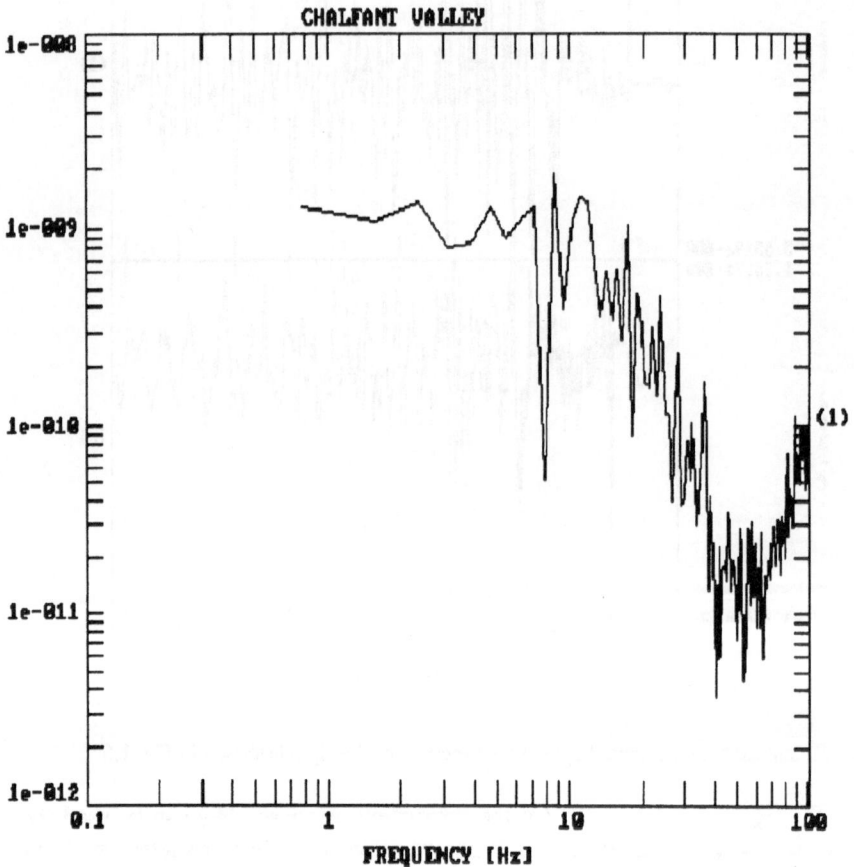

Fig. 1.8 Displacement spectrum for the P- wave portion of the instrument corrected displacement record of station 204 (top trace in Fig. 1.7).

Under very general assumptions, the displacement spectrum of an observed seismic P or S wave signal can be modelled as the product of different spectral factors:

$$S_j(f) = A(f) \cdot I_j(f) \cdot R_j(f) \cdot B_j(f) \qquad (1.1)$$

Here $S_j(f)$ represents the observed spectrum with j being the station index. On the right hand side, the contributing factors are the far field source spectrum $A(f)$, the instrumental response $I_j(f)$, the site response $R_j(f)$, and the attenuation spectrum $B_j(f)$. The far field displacement source spectrum is commonly modelled as:

$$A(f) = \frac{M_0 \cdot R(\theta, \phi)}{4 \cdot \pi \cdot \rho \cdot s \cdot v^3} \cdot \frac{f_c^\gamma}{f_c^\gamma + f^\gamma} \tag{1.2}$$

Here M_0 denotes the seismic moment, $R(\theta, \phi)$ the radiation pattern, ρ the density, s the hypocentral distance, v the P-, or S- wave velocity, respectively, f_c the source corner frequency, γ the high frequency decay factor (assumed to be 2 or 3 for P waves), and f the frequency, respectively. The attenuation spectrum $B(f)$ is given by:

$$B(f) = e^{-\pi f s/(vQ)} = e^{-\pi f t/Q} \tag{1.3}$$

with t being the traveltime. Once the instrument response has been removed equation (1.1) reduces to

$$S_j(f) = A(f) \cdot R_j(f) \cdot B_j(f) \tag{1.4}$$

If we know or make reasonable assumptions about the attenuation factor Q, the elastic parameters of the medium, and the site spectrum $R_j(f)$, equation (1.1) could directly be used to invert the spectrum in Fig. 1.8 for seismic source parameters. Inverse theory, however, is beyond the scope of this text. Readers interested in the topic of inversion are referred to books by Menke (1984), Tarantola (1987) and Hjelt(1992).

In order to understand the power as well as the limitations of the methods used for processing digital seismic data in the context of analysis and interpretation, we have to get a basic understanding of the principles of signal processing, or in other words, the theory of filters and systems. We will do this in a very practical way by starting out with trying to get a basic feeling for the underlying concepts. These concepts are - as we can see from the following definition - very general and not restricted to seismic signals.

> • *Definition* — Filters or systems are, in the most general sense, devices (in the physical world) or algorithms (in the mathematical world) which act on some input signal to produce a - possibly different - output signal.

Systems are often visualized in of block diagrams with a line or an arrow representing the input signal, and another representing the output signal (Fig. 1.9).

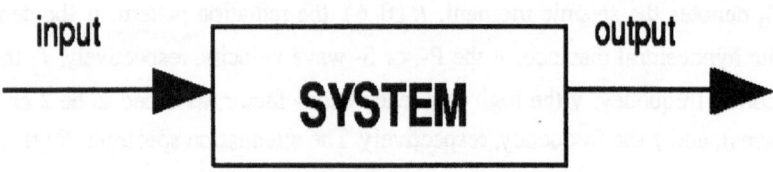

Fig. 1.9 Block diagram of a system

Using the concept of systems to model a seismic signal under very simplified assumptions, a possible way of representation is shown in Fig. 1.10. As seismologists, we are especially interested in isolating the left half of the diagram for interpretation in terms of geophysical models.

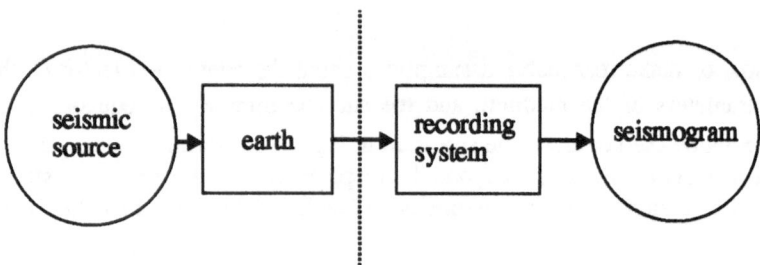

Fig. 1.10 System diagram of a seismogram

So far, the concept of a system is of little help, if it is only used to display the logical structure of a sequence of physical processes. In order to make real use of this concept, we first have to learn how to quantitatively describe the properties of systems and the way they act on signals. If a system meets some simple requirements (linearity, time invariance), we will see that we do not need to know what is physically going on inside of the system box. All we need to know in order to describe the properties of such a system and how it changes an arbitrary input signal is to know how it changes some very simple input signals like impulses or sinusoids. It will turn out that systems are most conveniently described in the so-called 'frequency' domain which is related to the time domain via the Fourier- or Laplace-transform.

2 RC Filter

We will start out with an example of a very simple filter which will enable us to know some of the basics from the theory of systems. The filter consists of a combination of a capacitor C and a resistor R in series (Fig. 2.1).

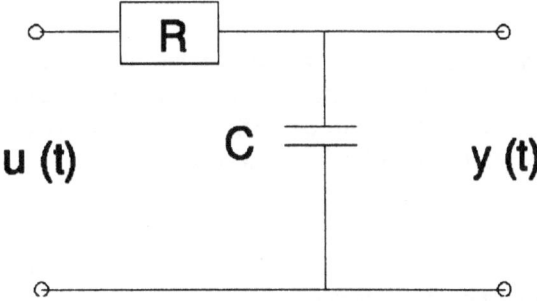

Fig. 2.1 RC filter.

It is easy to understand and allows us to compare different ways of quantifying the properties of this system (physical and non-physical one).

2.1 The system diagram

Making use of the little knowledge we have so far about the theory of filters, let us start out with the graphical representation of the RC filter as a simple block diagram (Fig. 2.2). All this tells us is that for the input signal $u(t)$, we obtain an output signal which we call $y(t)$.

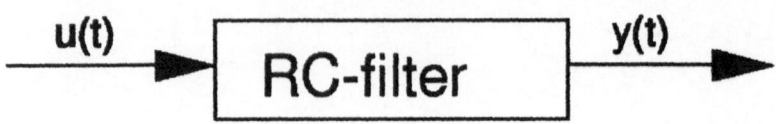

u(t) → RC-filter → y(t)

Fig. 2.2 System diagram for the RC filter in Fig. 2.1.

2.2 The differential equation

If we apply a time dependent input voltage $u(t)$ to the left outlets of the circuit, a current $I(t)$ will flow through the resistor R and the capacitor C. The voltage difference measured across the resistor will be $RI(t)$. If we call the voltage at the two right pins $y(t)$ we get for the overall voltage balance:

$$RI(t) + y(t) = u(t) \tag{2.1}$$

The current is controlled by the capacitance C:

$$I(t) = C\dot{y}(t) \tag{2.2}$$

Inserting equation (2.2) into equation (2.1) we obtain the differential equation of the electric circuit:

$$RC\dot{y}(t) + y(t) - u(t) = 0 \tag{2.3}$$

We have just arrived at a physical way to describe the properties of this circuit by a differential equation. Equation (2.3) is an example of a first order linear differential equation. For that reason, we call the corresponding system a **linear system**. Linear systems have the property that if $y_1(t), y_2(t)$ are the output signals corresponding to the input signals $x_1(t), x_2(t)$, respectively, the input signal

$$x_3(t) = \alpha_1 x_1(t) + \alpha_2 x_2(t) \tag{2.4}$$

will produce the output signal:

$$y_3(t) = \alpha_1 y_1(t) + \alpha_2 y_2(t) \tag{2.5}$$

We would also call this system **time invariant** since the properties of the filter (as described by its resistor and the capacitor) are assumed to be constant in time.

2.3 The frequency response function

For a zero input signal $(u(t) = 0)$ we obtain the homogeneous equation $RC\dot{y}(t) + y(t) = 0$. We can easily see by restitution that $y(t) = -\dfrac{1}{RC} e^{-t/(RC)}$ is a solution. However, we want to know how our filter acts on any kind of input signals. Arbitrary functions can be described as a superposition of harmonics under very general conditions (Fourier series or Fourier integrals). These powerful concepts will be used quantitatively in some detail in later chapters. At this point, a qualitative understanding of a few essential properties will be completely sufficient. We can think of the Fourier transform of mapping a 'time domain' signal into a 'frequency domain' signal which is called the corresponding spectrum. The actual transformation is done by projecting the signal to be transformed onto general harmonic functions (complex exponential functions) which can be thought of as some sort of coordinate system in a function space. In other words, the spectrum can be thought of representing the components of a 'time signal' in terms of harmonic functions. In the context of filtering, this provides a very convenient way to describe general linear systems. We can obtain the output signal corresponding to an arbitrary input signal by considering the output signals for harmonic input signal $u(t) = A_i e^{j\omega t}$ and then superimpose the responses for the individual frequencies. This would correspond to the calculation of the inverse Fourier transform. To solve equation (2.3) for a general harmonic signal, we make the classical Ansatz for the output signal:

$$y(t) = A_0 e^{j\omega t} \tag{2.6}$$

$$\dot{y}(t) = j\omega A_0 e^{j\omega t} \tag{2.7}$$

Inserting equations (2.6) and (2.7) into equation (2.3) we obtain:

$$A_0 e^{j\omega t} (RCj\omega + 1) = A_i e^{j\omega t} \tag{2.8}$$

and

$$\frac{A_0}{A_i} = \frac{1}{RCj\omega + 1} = U(j\omega) \tag{2.9}$$

$U(j\omega)$ is called the **frequency response function**. It is a complex quantity and we can separate it in its polar form using $1/(\alpha + j\beta) = \alpha/(\alpha^2 + \beta^2) - j\beta/(\alpha^2 + \beta^2)$ and $\phi = \text{atan}(Im/Re)$:

$$U(j\omega) = \frac{1}{\sqrt{1 + (RC\omega)^2}} e^{j\Phi} \tag{2.10}$$

$$\Phi = \text{atan}(-\omega RC) = -\text{atan}(\omega RC)$$

For an harmonic input signal with frequency ω, the phase shift depends on frequency as well as on the product of capacitance and resistance. Some very general lessons can be learned from this example so far:

• *The frequency response values are the complex eigenvalues of the system* — The output of the filter for an harmonic input signal is again a harmonic signal with different amplitude and phase. Therefore, the values of the frequency response function are the eigenvalues of the system (cf. $A\dot{x} = \lambda\dot{x}, \dot{x} \Leftrightarrow$ input signal; $A\dot{x} \Leftrightarrow$ output signal .

- *The frequency response is valid for arbitrary input signals* — If $A_i(j\omega)$ is the harmonic component of an arbitrary input signal, $A_o(j\omega)$ becomes the corresponding harmonic component of the output signal. Hence, the frequency response function relates the Fourier spectrum of the output signal $A_o(j\omega)$, to the Fourier spectrum of the input signal $A_i(j\omega)$:

$$U(j\omega) \; = \; \frac{A_o(j\omega)}{A_i(j\omega)} \tag{2.11}$$

For an arbitrary linear filter, the frequency response function uniquely describes its properties. The spectrum of the filter output signal is obtained by multiplying (complex multiplication) the spectrum of the input signal by the frequency response function of the filter.

- *The frequency response function is the Fourier transform of the impulse response* — For an impulsive input signal $u_g(t) = \delta(t)$, the spectrum is $A_i(j\omega) = 1$ and the output spectrum becomes $A_o(j\omega) = U(j\omega)$. In other words, the frequency response function is the Fourier transform of the impulse response of the system. The multiplication of the spectrum of the input signal with the frequency response function of the filter is equivalent to convolving the input signal with the impulse response of the filter. This property of linear systems is described by the *convolution theorem* and is discussed in some detail in chapters <pitsa>7.1 Fourier transform of continuous-time signals and <pitsa>7.5 The Discrete Fourier Transform (DFT). Here it is simply stated as an essential property of linear systems which will allow us to apply a filter operation either in the 'time domain' or in the 'frequency domain'.

The frequency response function is an extremely important tool in signal processing. It can be measured by comparing output and input signals to the system without further knowledge of the physics going on inside the filter. Once we know the frequency response function of a filter, we can predict the output of a filter to an arbitrary input signal.

2.4 The transfer function

The frequency response function is closely related to the concept of the transfer function. To show that let us solve equation (2.3) again by using the Laplace transformation. The bilateral Laplace transform of a function $f(t)$ is defined as:

$$L[f(t)] = \int_{-\infty}^{\infty} f(t) e^{-st} dt \tag{2.12}$$

with the complex variable $s = \sigma + j\omega$. $L[f(t)]$ will be written as $F(s)$. An important property of the Laplace transform in the context of solving differential equations is that the derivative in the time domain corresponds to a multiplication with s in the Laplace domain (which is also called the complex s plane):

$$L[\dot{f}(t)] = sF(s) \tag{2.13}$$

Transforming equation (2-3), we obtain:

$$RCsY(s) + Y(s) - U(s) = 0 \tag{2.14}$$

with $Y(s)$ and $U(s)$ being the Laplace transforms of $y(t)$ and $u(t)$, respectively.

> • *Definition* — The transfer function $T(s)$ is defined as the Laplace transform of the output signal divided by the Laplace transform of the input signal:

$$T(s) = \frac{Y(s)}{U(s)} = \frac{1}{1 + sRC} = \frac{1}{1 + s\tau} \tag{2.15}$$

If we set $s = j\omega$ we obtain equation (2.9), the frequency response function. This is due to the fact that the Fourier transform equals the Laplace transform evaluated along the imaginary axis of the s plane as we can see by replacing s by $j\omega$ in equation (2.12). In other words, the frequency response function can be defined as the Fourier transform of the output signal divided by the Fourier transform of the input signal. In signal processing literature, the term transfer function is sometimes used for the frequency response function as well. Although this is unfortunate, it should become clear from the context.

If we look at equation (2.15), we can see that $T(s)$ grows beyond limits for $s = -1/\tau$. It is said, that $T(s)$ has a pole at this location. We will see in the following that the existence and the position of the pole at $s = -1/\tau$ are sufficient to describe most the properties of the transfer function.

2.5 The impulse response

The transfer function given in equation (2.15) can be written as $F(s) = K/(s+a)$ with $C = 1/\tau$ and $a = 1/\tau$. Considering the signal $f(t) = e^{-at}h(t)$, with $h(t)$ being the unit step function (e.g. Strum and Kirk, 1988), we obtain its Laplace transform from (2.12) as

$$F(s) = \int_{-\infty}^{\infty} e^{-at}e^{-st}h(t)\,dt = \int_{0}^{\infty} e^{-(s+a)t}\,dt = -\left.\frac{e^{-(s+a)t}}{s+a}\right|_{0}^{\infty} \qquad (2.16)$$

We can see right away that (2.16) exists only for Re{s+a} > 0 or Re{s} > -a where the result becomes $1/(s+a)$. Hence, equation (2.15) is the Laplace transform of

$$y(t) = \frac{1}{\tau}e^{-\frac{1}{\tau}t} \qquad \text{for } t > 0 \qquad (2.17)$$

In the context of introducing the frequency response it was stated that the frequency response function is the Fourier transform of the impulse response of a system. With respect to the Laplace transform, we can make a similar statement. *The transfer function of a system is the Laplace transform of its impulse response function.* Hence, equation (2.17) describes the response of the RC filter to an impulsive input voltage.

The region where (2.16) exists is called region of convergence of the Laplace transform and in this case becomes the right half plane with Re {s} > -1/τ as shown by the shaded region in Fig. 2.3.

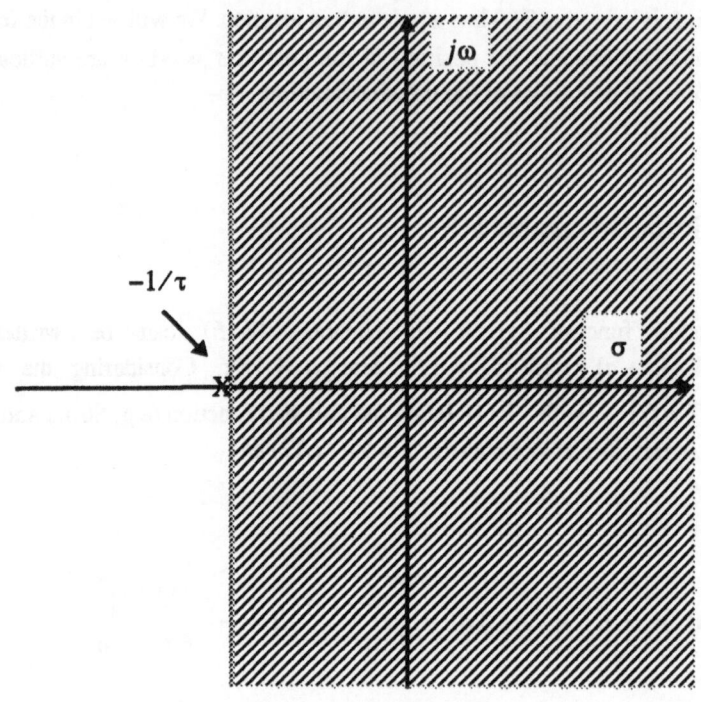

Fig. 2.3 Region of convergence of (2.16). The pole location at $-1/\tau$ is marked by an X.

Since we are using the bilateral Laplace transform, it should be noted, that the Laplace transform of $f(t) = -e^{-at}h(-t)$, with $h(-t)$ being the time inverted unit step function:

$$F(s) = -\int_{-\infty}^{\infty} e^{-at}e^{-st}h(-t)\,dt = -\int_{-\infty}^{0} e^{-(s+a)t}dt = \left.\frac{e^{-(s+a)t}}{s+a}\right|_{-\infty}^{0} \qquad (2.18)$$

also comes out to be $1/(s+a)$. However, we see that it exists only for Re{s+a} < 0 or Re{s} < -a which corresponds to the shaded region in Fig. 2.4.

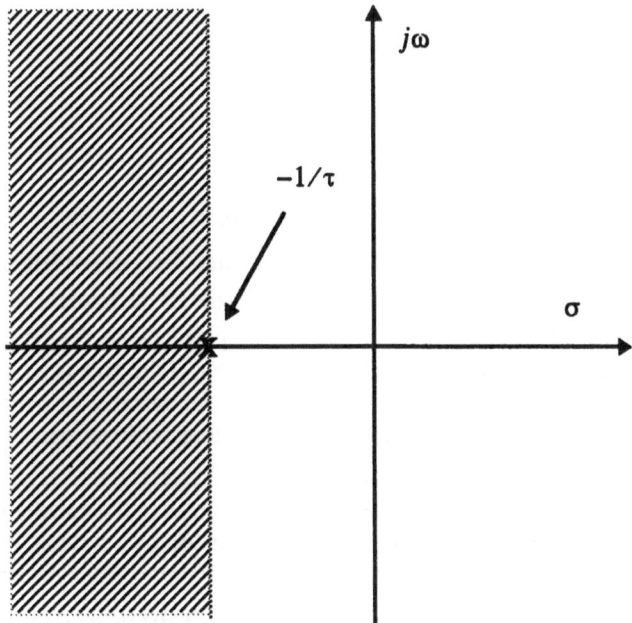

Fig. 2.4 Region of convergence of (2.18). The pole location at $-1/\tau$ is marked by an X.

When we formally calculate the impulse response function from the transfer function we have to calculate the inverse Laplace transform, defined as

$$\mathcal{L}^{-1}[F(s)] = f(t) = \frac{1}{2\pi j} \int_{\sigma - j\infty}^{\sigma + j\infty} F(s)\, e^{st} ds \qquad (2.19)$$

The path of integration must lie in the region of convergence. If we evaluate (2.19) on the imaginary axis, which means for $s = j\omega$, (2.19) becomes the equation of the inverse Fourier transform. Since for $s = j\omega$ the transfer function becomes the frequency response function, this means that the impulse response can either be calculated by inverse Fourier transform from the frequency response function or by inverse Laplace transform from the transfer function.

Depending on whether the region of convergence which is going to be considered for the evaluation of (2.19) is a right half plane or a left half plane we will get a right-sided or a left-sided impulse response function, respectively. For the given example $F(s) = 1/(s+a)$, the right sided function $f(t) = e^{-at}h(t)$ cor-

responds to a causal system ($f(t) = 0$ for $t < 0$) while the left-sided function
$f(t) = -e^{-at}h(-t)$ corresponds to an anti-causal signal which vanishes for
$t > 0$ and is physically not realizable.

2.5.1 The condition for stability

The physically realizable impulse response of the RC filter
$[y(t) = (1/\tau)e^{(-1/\tau)t}h(t)]$ has an exponential time dependence with the
exponent $(-1/\tau)$ being exactly the location of the pole of the transfer function. As
long as the pole is located in the left half of the complex s plane, the causal impulse
response will decay exponentially with time. However, if the pole is located within
the right half plane, the impulse response will become unstable (growing beyond
limits). This rule is valid also for more complicated transfer functions:

> In order for a causal system to be stable, all the poles of the transfer func-
> tion have to be located within the left half of the complex s plane.

It should be noted however, that for anticausal signals the opposite is true. For a
pole at $1/\tau$, the anticausal signal $y(-t) = (1/\tau)e^{(1/\tau)(-t)}h(-t)$ would well
be stable, although physically unrealizable.

2.6 The frequency response function and the pole position

Given the pole location of the transfer function on the complex s plane we can
determine the frequency response function of the system using a simple graphical
method (Fig. 2.5).

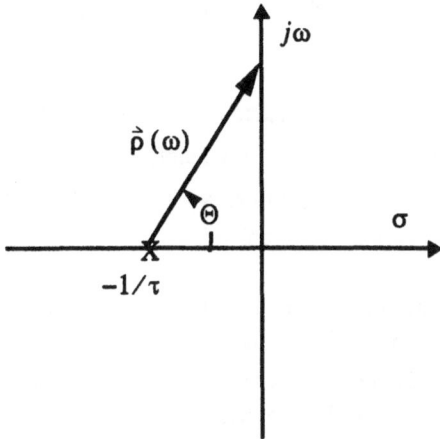

Fig. 2.5 Representation of the RC filter in the s plane. The pole location at $-1/\tau$ is marked by an X.

For $s = j\omega$, ω moves along the imaginary axis. Rewriting equation (2.15) we get:

$$T(s) = \frac{1}{1+s\tau} = \frac{1}{\tau}\left[\frac{1}{\frac{1}{\tau}+s}\right] \tag{2.20}$$

and with $s = j\omega$

$$T(j\omega) = \frac{1}{\tau}\left[\frac{1}{\frac{1}{\tau}+j\omega}\right] \tag{2.21}$$

Written as a complex number, $1/\tau + j\omega$ represents the pole vector $\vec{\rho}(\omega)$ which is pointing from the pole position towards the actual frequency on the imaginary axis. Using $\vec{\rho}(\omega)$ and polar coordinates we obtain for the frequency response function:

$$T(j\omega) = \frac{1}{\tau}\left[\frac{1}{|\vec{\rho}(\omega)|e^{j\theta}}\right] = \frac{1}{\tau}\left[\frac{1}{|\vec{\rho}(\omega)|}e^{-j\theta}\right] = |T(j\omega)|e^{j\Phi} \quad (2.22)$$

For the given example, the amplitude value of the frequency response function for frequency ω is inversely proportional to the length of the pole vector $\vec{\rho}(\omega)$ from the pole location to the point $j\omega$ on the imaginary axis. The phase angle equals the negative angle between $\vec{\rho}(\omega)$ and the real axis.

Problem 2.1 Determine graphically the amplitude characteristics of the frequency response for a RC filter with $R = 4.0\Omega$ and $C = 1.25F/2\pi = 0.1989495F$ $(1\Omega = 1(V/A), 1F = 1A\sec/V)$. Where is the pole position in the s plane? For the plot use frequencies between 0 and 5 Hz.

The graphical way to determine the frequency response function is quite instructive, since it provides a quick look at the system's properties in the frequency domain. However, for a quantitative analysis we would of course determine $T(j\omega)$ by directly evaluating equation (2.21) for different frequency values. The result of solving Problem 2.1 numerically using PITSA is shown in Fig. 2.6.

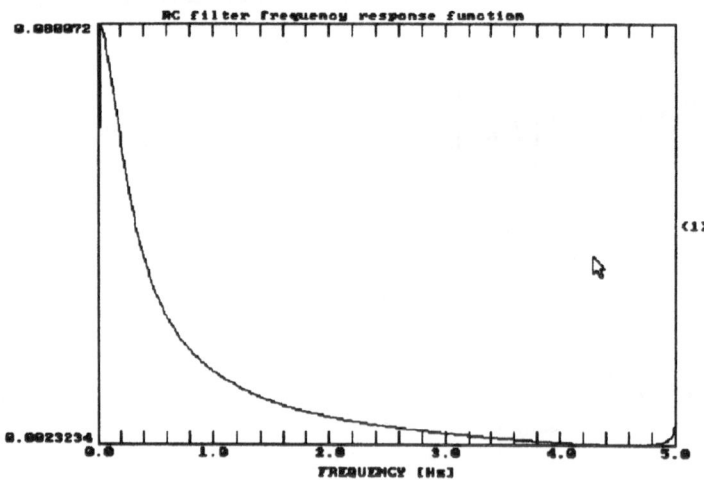

Fig. 2.6 Frequency response function (amplitude only) of the RC filter of example 2.1

We see that for high frequencies, the amplitude values of the frequency response function decrease continuously. We could have guessed this already (or known from physics class) since for low frequencies, the capacitor will act like an resistor of infinite resistance, while for high frequencies it will act like a short circuit. Therefore, the high frequency components of the input signal will not make it to the output signal. In other words, the circuit acts as a low pass filter.

Problem 2.2 Calculate the frequency response of the RC filter from problem 2.1 using PITSA.

We can learn some more about the effect of the pole on the frequency response function by displaying the same amplitude spectrum again in a log - log scale (Fig. 2.7).

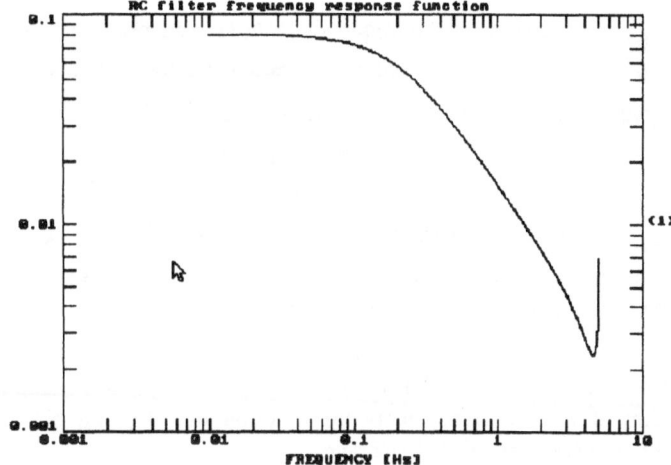

Fig. 2.7 Same plot as Fig. 2.6 only on a log-log scale

In the log-log plot, we can see that the amplitude portion of the frequency response function can be approximated in the high- and low frequency limit roughly by two straight lines. These lines intersect at a frequency of .2 Hz which is called the corner frequency of the filter. In the case of the RC filter, it is equal to $1/RC = 1/(5 \text{ sec})$ which turned out also to be the exponential term in the impulse response.

We can understand this more quantitatively by going back to equation (2.21) and concentrating on the amplitude portion of it.

$$|T(j\omega)| = \frac{1}{\tau}\left[\frac{1}{\left|\frac{1}{\tau} + j\omega\right|}\right] = \frac{1}{\sqrt{1 + \omega^2\tau^2}} \qquad (2.23)$$

If we define $|-1/\tau|$, the distance of the pole position from the origin of the s plane, as ω_c, we get

$$|T(j\omega)| = \frac{1}{\sqrt{1 + \dfrac{\omega^2}{\omega_c^2}}} \qquad\qquad (2.24)$$

For $\omega \to 0$, $|T(j\omega)| \to 1 = const$ and for $\omega \gg \omega_c$ we easily see that $|T(j\omega)| \approx \omega^{-1}$. Thus, the slope of the frequency response function (amplitude portion) in a log-log plot changes from being 0 for frequencies much smaller than ω_c to -1 for frequencies much larger than ω_c. If we measure the slope in dB (decibel) - which is defined from the amplitude ratio of two reference signals A_{max} and A_{min} - we obtain $Slope_{dB} = 20\log_{10}(A_{max}/A_{min})$. A_{max} and A_{min} are normally taken either a decade (factor 10) or an octave (factor 2) apart in frequency. A slope of -1 corresponds to 20 dB/decade and 6 dB/octave, respectively. We can state the general rule:

> Rule: A single pole in the transfer function causes the slope of the amplitude portion of the frequency response function in a log-log plot to decrease by 20 dB/decade or 6 dB/octave.

We can also apply this rule the other way round. From the decrease of the slope of the amplitude portion of a given frequency response function at a corner frequency ω_c, we can suspect a pole to exist which has to be located at a distance $|\omega_c|$ away from the origin of the s plane. We will see later how this rule can be extended to zeroes of the transfer function and to more complicate transfer functions as well.

2.7 The difference equation

Yet another method to represent systems, extensively used in processing discrete data, is by difference equations. For the RC circuit, we arrive at such a representation by approximating the derivative at time t in equation (2.3) by a finite difference at time nT, with the discretization interval being T.

$$\dot{y}(t) = \frac{dy(t)}{dt} = \frac{dy(nT)}{dt} \approx \frac{y(nT) - y((n-1)T)}{T} \qquad (2.25)$$

Equation (2.3) becomes

$$RC\frac{y(nT) - y((n-1)T)}{T} + y(nT) = u(nT) \qquad (2.26)$$

Writing $y(nT)$ as $y(n)$ this leads to

$$y(n) = b_0 u(n) - a_1 y(n-1) \qquad (2.27)$$

with

$$b_0 = \frac{\dfrac{T}{RC}}{1 + \dfrac{T}{RC}}$$

$$\qquad (2.28)$$

$$a_1 = -\frac{1}{1 + \dfrac{T}{RC}}$$

The output signal $y(n)$ at time nT depends on the value of the input signal $u(n)$ at time nT as well as on the value of the output signal at time $(n-1)T$. Equation (2.27) can be solved for arbitrary input signals by numerical recursion.

Problem 2.3 Let us end this chapter by considering an example directly related to our daily life. Consider a savings account with a monthly interest rate of α percent. The money which is deposited at time $t = nT$ is supposed to be $x(nT)$. $y(nT)$ represents the money in the account at time nT (before the deposit of $x(nT)$ is made), and $y(nT + T)$ is the money one sample (1 month) later. Determine the difference- and differential equations of the system using the forward difference $(\dot{y}(t) \approx \dfrac{y(nT+T) - y(nT)}{T})$. Start out with the balance at time $t = nT + T$ which can be written as $y(nT + T) = y(nT) + \alpha y(nT) + \alpha x(nT) + x(nT)$. Calculate the transfer function using Laplace transform (use equation (2.13)). Is the system stable. Could we use an RC filter to simulate the savings account?

2.8 Review

The central theme of this chapter was to study the behaviour of a simple electric RC circuit. We introduced the term filter or system as a device or algorithm which changes some input signal into an output signal. We saw that the RC filter is an example for a **linear, time invariant (LTI) system**, which could be described by a linear differential equation. From the solution of the **differential equation** for harmonic input we obtained the result that the output is again a harmonic signal. We introduced the concept of the **frequency response function** as the Fourier transform of the output signal divided by the Fourier transform of the input signal. The frequency response function was seen to have important properties:

• The values of the frequency response function are the eigenvalues of the system.

• Knowing the frequency response function, we can calculate the output of the filter to arbitrary input signals by multiplying the Fourier transform of the input signal with the frequency response function.

• The frequency response function is the Fourier transform of the impulse response. Knowing the impulse response function, we can calculate the output of the filter to arbitrary input signals by convolving the input signal with the impulse response function.

We then introduced the concept of the **transfer function** as an even more general concept to describe a system as the Laplace transform of the output signal divided by the Laplace transform of the input signal. The transfer function   can also be seen as the Laplace transform of the impulse response function. The frequency response function could be derived from the transfer function by letting $s = j\omega$. We found that the transfer function of the RC circuit has a pole at the location $-1/RC$ (on the negative real axis of the s plane). We also found that the (causal) impulse response of a system with a single pole is proportional to an exponential function e^{pt} with p being the location of the pole. Therefore the causal system can only be **stable** if the pole is located within the left half plane of the s plane. We found a way to graphically determine the frequency response function given the pole position in the s plane. From analysing the frequency response function in a log-log plot, we derived the rule that a pole in the transfer function causes a change of the slope of the frequency response function at a frequency ω_c by 20 dB/decade with ω_c being the distance of the pole from the origin of the s plane. We finally approximated the **differential equation** of the RC circuit by its difference equation which could be solved iteratively.

3 General linear time invariant systems

3.1 Generalization of concepts

The reason for studying the simple RC circuit in as much detail as we did was that the concepts we used for its analysis stay valid for much more complicated systems. In the following we will assume a general system with the only restriction of it being linear and time invariant (LTI system). We will start out by looking at the differences that makes to our tools for the analysis of systems. Table 3.1 shows how the different concepts we have been using to describe the RC-filter have to be formally extended in order to describe general LTI systems. For example, if we rewrite the differential equation for the RC filter (2.3):

$$RC\dot{y}(t) + y(t) - x(t) = \alpha_1 \frac{d}{dt} y(t) + \alpha_0 y(t) + \beta_0 x(t) = 0 \qquad (3.1)$$

we can see it as a special case of an Nth order LTI system

$$\sum_{k=0}^{N} \alpha_k \frac{d^k}{dt^k} y(t) + \sum_{k=0}^{L} \beta_k \frac{d^k}{dt^k} x(t) = 0 . \qquad (3.2)$$

Similarly, we can see in Table 3.1 how the other concepts are extended to the general case.

In chapter 2.5 we have seen that for a system with a single pole, the impulse response could be left-sided or right-sided depending on the region of convergence considered. For a general LTI system, the region of convergence consists of bands parallel to the imaginary ($j\omega$) axis which do not contain any poles of the transfer function.

Table 3.1 Correspondences between the RC filter (1st order system) and a general Nth order LTI system

Concept	RC filter	General system
Differential equation	$\alpha_1 \dfrac{d}{dt} y(t) + \alpha_0 y(t)$ $+ \beta_0 x(t) = 0$	$\displaystyle\sum_{k=0}^{N} \alpha_k \dfrac{d^k}{dt} y(t) + \sum_{k=0}^{L} \beta_k \dfrac{d^k}{dt} x(t) = 0$
Transfer function	$T(s) = -\dfrac{\beta_0}{\alpha_0 + \alpha_1 s}$	$T(s) = -\dfrac{\beta_0 + \beta_1 s + \beta_2 s^2 + \dots + \beta_L s^L}{\alpha_0 + \alpha_1 s + \alpha_2 s^2 + \dots + \alpha_N s^N}$
Frequency response function	$T(j\omega) = -\dfrac{\beta_0}{\alpha_0 + \alpha_1 j\omega}$	$T(j\omega) = -\dfrac{\beta_0 + \beta_1 (j\omega) + \beta_2 (j\omega)^2 + \dots + \beta_L (j\omega)^L}{\alpha_0 + \alpha_1 (j\omega) + \alpha_2 (j\omega)^2 + \dots + \alpha_N (j\omega)^N}$
Poles and zeroes	A single pole at the root of the denominator polynomial	N poles at the roots of the denominator polynomial, L zeroes at the roots of the numerator polynomial
Difference equation	$y(n) = -a_1 y(n-1)$ $+ b_0 x(n)$	$y(n) = -\displaystyle\sum_{k=1}^{N} a_k y(n-k) + \sum_{k=0}^{L} b_k x(n-k)$

For a right-sided signal, the region of convergence is always a right half plane, while for a left-sided signal it is a left half plane. For a two sided signal it consists of a finite band. For the general LTI systems which will be considered here, the transfer functions are always rational functions. In this case the regions of convergence are always bounded by poles. Finally, for a signal of finite duration the region of convergence is always the complete s plane which means it cannot be represented by poles an zeroes, an issue of importance at a later state. Fig. 3.1 shows the different types of convergence regions for a general LTI system with an existing Laplace transform and the corresponding types of stable impulse response functions (IR). A system which does not belong to any of these four classes does not have a Laplace transform at all.

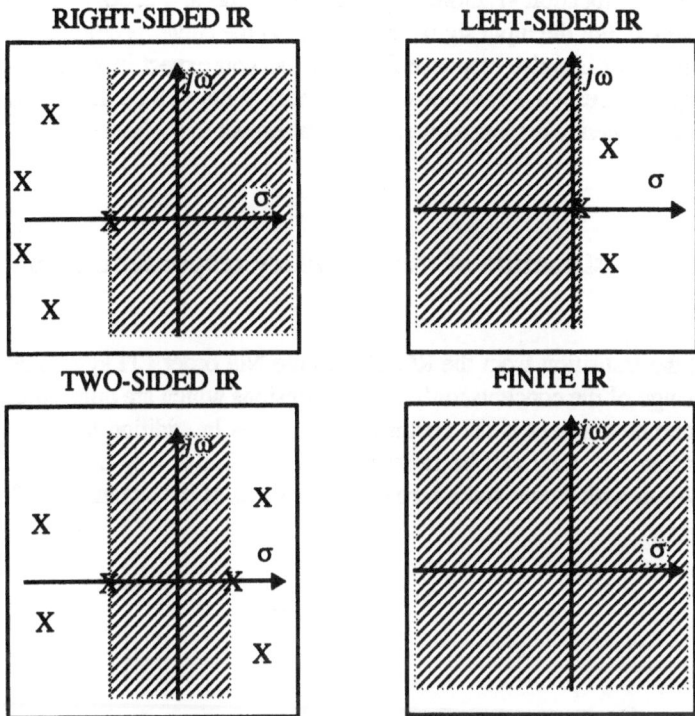

Fig. 3.1 Types of convergence regions for general LTI systems with existing Laplace transform and the corresponding types of stable (infinite duration) impulse response (IR) functions.

When an impulse response function is calculated numerically for a given pole-zero distribution, commonly the inverse Fourier transform is used (actually the discrete Fourier transform, but we can ignore the differences for the following argument). Thus the path of integration is fixed to the imaginary axis. The type of impulse response calculated will therefore be the one whose region of convergence contains the imaginary axis.

Problem 3.1 Change the pole-zero distribution from Problem 2.2 to describe a system with two poles. Consider three different cases. a) Put both poles at -1.2566, 0. b) Put one pole at location -1.2566, 0 and the other one at 1.2566, 0. c) Put both poles at 1.2566, 0. For the input signal, use a spike at the center position of the window. What types of impulse response functions do you expect in all three cases? Will the frequency response functions be different? What will be the changes you expect for the frequency response functions with respect to Problem 2.2?

3.2 Graphical estimation of the frequency response function

We see that the transition from the RC filter to the Nth order LTI system does not require a change of the concepts, only some extensions which are quite straightforward. The one new aspect is the occurrence of zeroes in addition to poles. However, as it turns out, they can be treated in a very similar way to poles. Let us assume a system with a pole and a zero on the real axis of the s plane (Fig. 3.2)

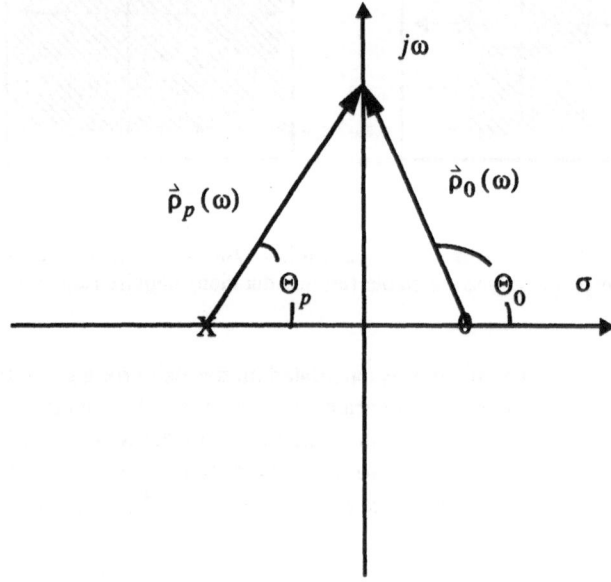

Fig. 3.2 Complex s plane representation of a system with a single pole and zero. The pole and zero locations are marked by an X, and a 0, respectively.

The transfer function for this case becomes:

$$T(s) = \frac{s - s_0}{s - s_p} \tag{3.3}$$

with s_0 and s_p being the position of the zero and the pole, respectively. For the frequency response function ($s = j\omega$) we get

$$T(j\omega) = \frac{j\omega - s_0}{j\omega - s_p} \tag{3.4}$$

Written as a complex number, $j\omega - s_p$ and $j\omega - s_0$ represent the vectors $\vec{p}_p(\omega)$ and $\vec{p}_0(\omega)$ which are pointing from the pole position and the zero position, respectively, towards the actual frequency on the imaginary axis. For the frequency response function we obtain:

$${}^r(j\omega) = |\vec{p}_o(\omega)| e^{j\theta_0} \cdot \frac{1}{|\vec{p}_p(\omega)|} e^{-j\theta_p} \tag{3.5}$$

The amplitude value of the frequency response function for frequency ω equals the length of the vector $\vec{p}_0(\omega)$ from the zero location to the point $j\omega$ on the imaginary axis divided by the length of the vector $\vec{p}_p(\omega)$ from the pole location to the point $j\omega$ on the imaginary axis. The phase angle equals the angle between $\vec{p}_o(\omega)$ and the real axis minus the angle between $\vec{p}_p(\omega)$ and the real axis.

Extrapolating this example, we obtain a graphical method for the determination of the frequency response function of an arbitrary LTI system:

> The amplitude part of the frequency response function of an arbitrary LTI system can be determined graphically by multiplying the lengths of the vectors from the zero locations in the S plane to the point $j\omega$ on the imaginary axis divided by the product of all lengths of vectors from pole locations to the point $j\omega$ on the imaginary axis. Likewise, to determine the phase part, the phase angles for the vectors from the zero locations in the S plane to the point $j\omega$ on the imaginary axis have to be added together. Then, the phase angles of all the vectors from pole locations to the point $j\omega$ on the imaginary axis have to be subtracted.

Problem 3.2 Use the argument given above to determine the frequency response for Problem 2.2 if you add a zero at position 1.2566, 0?

3.3 The phase properties of general LTI system

In the preceding discussion it has been shown that for the amplitude part of the frequency response only the lengths of the pole- and zero-vectors $\vec{\rho}_p(\omega)$ and $\vec{\rho}_o(\omega)$, respectively, have to be considered. For the example shown in Fig. 3.3 this means that because the zeroes in a) and b) appear as mirror images with respect to the imaginary axis, both systems will have the same amplitude response. For the phase response we can easily see that the angle between $\vec{\rho}_o(\omega)$ and the real axis will always be greater for a zero in the right half plane (Fig. 3.3b) than for a zero in the left half plane (Fig. 3.3a).

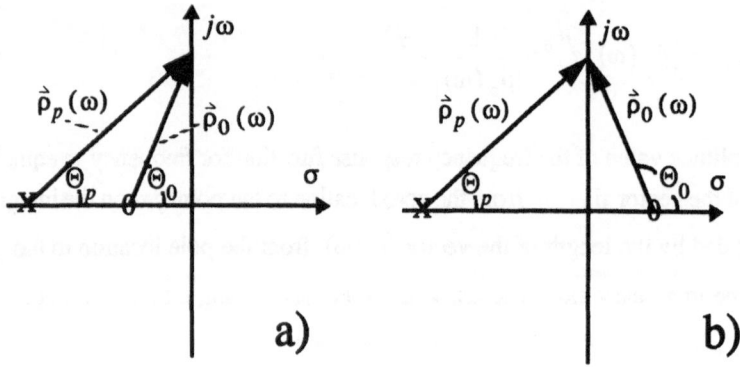

Fig. 3.3 Complex s plane representation of two systems with a single pole and zero having the same amplitude response. The pole and zero locations are marked by an X, and a 0, respectively.

For general LTI systems, zeroes in the right half plane will always yield larger phase response contributions than zeroes in the left half plane. For a given amplitude response, the phase response will have the smallest possible values if all zeroes are located in the left half plane. This leads us quite naturally to the concept of minimum and maximum phase:

> • *Definition* — A causal, stable system (no poles in the right half plane) is minimum phase provided it has no zeroes in the right hand plane. It is maximum phase if it has all its zeroes in the right hand plane.

Minimum phase systems have a number of desirable properties which are especially important in the context of digital filters. Systems which are neither minimum phase nor maximum phase are called mixed phase. If a filter performs no phase distortion but causes a constant time shift for all frequencies, its phase response must be directly proportional to frequency. This can be understood from the shifting property of the Fourier transform (see(7.4) on page 76). These types of filters are called linear phase. Filters for which have the phase response is zero for all frequencies are called zero phase filters. They can implicitely be created by filtering the same signal twice in opposite directions, thus cancelling their phase responses.

Problem 3.3 How can the following two statements be proven for a general LTI system? a) If a system is minimum phase it will always have a stable and causal inverse filter. b) Any mixed phase system can be seen as a convolution of a minimum phase system and an allpass filter, which only changes the phase response but leaves the amplitude response as is.

Problem 3.4 How can we change the two-sided impulse response from Problem 3.1b into a right-sided one without changing the amplitude response? Keyword: allpass filter.

3.4 The interpretation of the frequency response function

From the interpretation of the RC circuit we had concluded that a single pole in the transfer function causes the slope of the amplitude portion of the frequency response function in a log-log plot to decrease by 20 dB/decade (6 dB/octave). The transition takes place at the frequency ω_c which has been found to be equal to the modulus of the pole position. If we take the inverse of the transfer function, a single pole will become a single zero and we can conclude likewise that a single zero in the transfer function causes the slope of the amplitude portion of the frequency response function in a log-log plot to increase by 20 dB/decade or 6 dB/octave. The transition takes place at a frequency ω_c which is equal to the modulus of the zero position.

Problem 3.5 Consider a system with a pole and a zero on the real axis of the s plane. Let the pole position be (-6.28318, 0), and the zero position (.628318,0). What is the contribution of the zero to the frequency response function?

Using the general rule above, we can directly interpret the shape of the amplitude part of the frequency response of a general LTI system in terms of the location of poles and zeroes. Multiple poles or zeroes will contribute slope changes by a multiple of 20 dB/decade.

Before we end the discussion of the effects of poles and zeroes on the transfer function, there is one more point to discuss. So far, we have only considered poles and zeroes on the real axis of the s plane. On the other hand, from the definition of the corner frequency ω_c, we have seen that it is only dependent on the distance of the singularity from the origin of the s plane. In other words, all poles located on a circle around the origin of the s plane will produce the same corner frequency ω_c in the frequency response function. So what is changing if we change the position of, let's say our single pole to along a circle? Let's check it out!

Just wait a second! Can we just shift a single pole along a circle? We have assumed so far that all our systems are linear time invariant systems. This means the systems could be described by linear differential equations. As long as the coefficients of the differential equations would be real quantities, however, the coefficients of the polynomials making up the transfer function would be real as well. The roots of a polynomial with real coefficients, however, can only be real or appear in complex conjugate pairs. So to answer our question: No, we cannot just shift a pole from the real axis. Poles and zeroes appearing off the real axis of the S plane have to appear in conjugate complex pairs.

Problem 3.6 Move the pole position of a double pole at (-1.2566, 0.) in steps of 15° (up to 75°, and 85°) along a circle centered at the origin and passing through the original double pole. Calculate the impulse response functions and the amplitude portions of the corresponding frequency response functions.

We can conclude that moving the poles along a circle around the origin of the s plane does not change the spectral roll off away from the corner frequency. However, it changes strongly the behaviour of the filter at the corner frequency. The closer the poles move towards the imaginary axis, the more do the impulse responses show 'resonance' effects.

Problem 3.7 Use the pole-zero approach to design a notch filter suppressing unwanted frequencies at 6.25 Hz. What kind of singularities do we need? How can we make use of the result of Problem 3.6?

Problem 3.8 From the shape of the frequency response function in Fig. 3.4, determine the poles and zeroes of the corresponding transfer function.

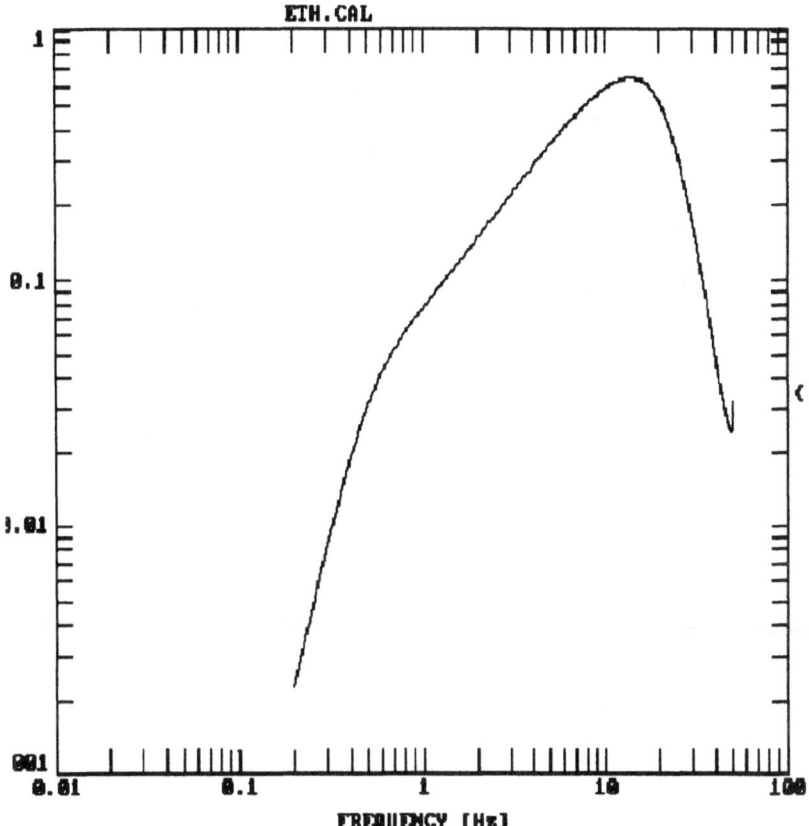

Fig. 3.4 Frequency response function (amplitude) with an unknown pole - zero distribution.

4 The seismometer

Up to this point we have encountered different methods to describe a linear time invariant system. We have learned to analyze and design simple filters in terms of poles and zeroes of the transfer function. We have also seen some of the interconnections between the different approaches.

Here we are now going to apply these concepts to the classical pendulum seismometer. From the analysis of the frequency response function and the transfer function of the seismometer we should get a better understanding of why and how seismic signals are changed when passing through a seismometer. This will bring us a step closer towards our final goal to possibly correct recorded seismograms for the effects of the recording process.

In order to understand how seismic signals are effected passing through a seismometer and how the output signal relates to the true ground motion, we will use the concepts of the last chapter to model a seismometer as a linear time invariant system. We will restrict ourselves to a simple vertical pendulum seismometer as it is schematically sketched in Fig. 4.1. It consists of at least three different elements tied together: A mass, which is connected to the frame by a spring and a damping mechanism. The frame is assumed to be firmly connected to the ground. What controls the actions of the seismometer is the superposition of all forces acting on the mass simultaneously:

• *The inertia of the mass* — The inertial force is acting on the moving mass m. It is directed into the opposite direction of the motion of the mass $u_m(t)$ (downwards if the ground moves upwards). It is most conveniently measured with respect to the inertial reference frame denoted as u.

$$f_i = -m\ddot{u}_m(t) \qquad\qquad (4.1)$$

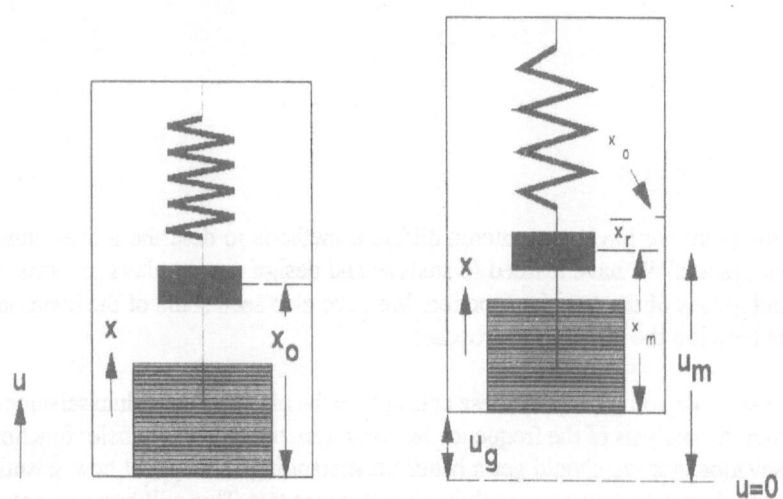

Fig. 4.1 Model of a vertical pendulum seismometer. The inertial coordinate system is denoted u, while the x coordinate system is moving with the frame.

• *The spring* — A second force f_{sp} will be exerted by the spring, if it is elongated. f_{sp} is proportional to the elongation ($x_r = x_m - x_0$ measured within the reference frame x which is attached to the seismometer frame). If the spring is elongated ($x_r < 0$), the force f_{sp} will be positive (pointing upwards).

$$f_{sp} = -kx_r(t)$$

k = spring constant (strength)   (4.2)

• *The dashpot* — Finally, the frictional force f_f is acting on the mass. It is proportional to the velocity $\dot{x}_m(t)$ with which the mass is moved with respect to the dashpot. As with the elongation of the spring, this quantity is best described within the reference frame x which is attached to the frame of the seismometer. If the movement is up, the frictional force will be directed downwards.

$$f_f = -D\dot{x}_m(t)$$

D = friction coefficient   (4.3)

In equilibrium, all these forces add up to zero:

$$-m\ddot{u}_m(t) - D\dot{x}_m(t) - kx_r(t) = 0 \tag{4.4}$$

Since $u_m(t) = u_g(t) + x_m(t)$, we obtain

$$-m(\ddot{u}_g(t) + \ddot{x}_m(t)) - D\dot{x}_m(t) - kx_r(t) = 0 \tag{4.5}$$

With $\dot{x}_m(t) = \dot{x}_r(t)$ and $\ddot{x}_m(t) = \ddot{x}_r(t)$ we get

$$m\ddot{x}_r(t) + D\dot{x}_r(t) + kx_r(t) = -m\ddot{u}_g(t) \tag{4.6}$$

Dividing by m we obtain the equation of motion for the seismometer

$$\ddot{x}_r(t) + \frac{D}{m}\dot{x}_r(t) + \frac{k}{m}x_r(t) = -\ddot{u}_g(t) \tag{4.7}$$

Rewriting the constant terms we can write

$$\ddot{x}_r(t) + 2\varepsilon\dot{x}_r(t) + \omega_o^2 x_r(t) = -\ddot{u}_g(t) \tag{4.8}$$

where $\omega_o^2 = \dfrac{k}{m}$ and $2\varepsilon = \dfrac{D}{m} = 2h\omega_0$. $h = \dfrac{\varepsilon}{\omega_0}$ is called the damping constant of the seismometer.

At this point we have already learned some important lessons:

— For slow movements, the $\ddot{x}_r$ and $\dot{x}_r$ become negligible and x_r dominates the left hand side of (4.8). This corresponds to the seismometer measuring ground acceleration $\ddot{u}_g$.

— For fast movements, the $\ddot{x}_r$ dominates the left hand side of (4.8) and the seismometer is measuring ground displacement.

In order to obtain the true motion of the ground for the general case, we have to calculate a weighted sum of the relative movement between the moving mass and the seismometer frame (x_r) and its first and second time derivative ($\dot{x}_r$ and $\ddot{x}_r$, respectively). Therefore, we need to know the damping factor ε or damping constant h, respectively, as well as the natural period of the seismometer ω_0.

4.1 The solution for simple initial conditions (release test)

One way to determine the damping factor and the natural period of a seismometer is by releasing the seismometer mass from a known starting displacement at time $t = 0$. (For an electrodynamic system this is normally done by applying a step in the current to the calibration coil). The corresponding initial conditions are:

$$x_r(0) = x_{r0}$$

(starting displacement)

$$\dot{x}_r(0) = 0$$

(mass is at rest at time $t = 0$)

$$\ddot{u}_g(t) = 0$$

(ground excitation is zero)

Equation ((4.8)) then becomes the homogeneous second order differential equation of the damped harmonic oscillator:

$$\ddot{x}_r(t) + 2\varepsilon\dot{x}_r(t) + \omega_0^2 x_r(t) = 0 \tag{4.9}$$

Making the classical Ansatz: $x_r(t) = Ae^{\alpha t}$ with $\dot{x}_r(t) = \alpha Ae^{\alpha t}$ and $\ddot{x}_r(t) = \alpha^2 Ae^{\alpha t}$, we obtain

$$(\alpha^2 + 2\varepsilon\alpha + \omega_0^2) Ae^{\alpha t} = 0 \tag{4.10}$$

Since $e^{\alpha t} \neq 0$ for all t we get

$$\alpha^2 + 2\varepsilon\alpha + \omega_o^2 = 0 \tag{4.11}$$

and the two solutions:

$$\alpha_1 = -\varepsilon + \sqrt{\varepsilon^2 - \omega_0^2}$$

$$\alpha_2 = -\varepsilon - \sqrt{\varepsilon^2 - \omega_0^2} \tag{4.12}$$

From the theory of linear differential equations we know that any linear combination of solutions is again a solution of the differential equation. Therefore, the general solution of equation (4.9) can be written as:

$$x_r(t) = A_1 e^{\alpha_1 t} + A_2 e^{\alpha_2 t} =$$

$$= A_1 e^{-\left(\varepsilon - \sqrt{\varepsilon^2 - \omega_0^2}\right)t} + A_2 e^{-\left(\varepsilon + \sqrt{\varepsilon^2 + \omega_0^2}\right)t} \tag{4.13}$$

The coefficients A_1 and A_2 have to be estimated from the initial conditions:

$$x_r(0) = x_{r0} = A_1 + A_2$$

$$\dot{x}_r(0) = 0 = \alpha_1 A_1 + \alpha_2 A_2 \tag{4.14}$$

We get

$$\alpha_1 A_1 = -\alpha_2 A_2 = -\alpha_2(x_{r0} - A_1) = -x_{r0}\alpha_2 + \alpha_2 A_1$$

$$A_1(\alpha_1 - \alpha_2) = -x_{r0}\alpha_2 \tag{4.15}$$

and the coefficients A_1 and A_2 become:

$$A_1 = x_{r0}\frac{\alpha_2}{\alpha_2 - \alpha_1} = x_{r0}\frac{\varepsilon + \sqrt{\varepsilon^2 - \omega_0^2}}{2\sqrt{\varepsilon^2 - \omega_0^2}}$$

$$A_2 = x_{r0}\frac{-\alpha_1}{\alpha_2 - \alpha_1} = x_{r0}\frac{-\varepsilon + \sqrt{\varepsilon^2 - \omega_0^2}}{2\sqrt{\varepsilon^2 - \omega_0^2}} \tag{4.16}$$

The solution of equation (4.9) finally becomes:

$$x_r(t) = \frac{x_{r0}}{2\sqrt{\varepsilon^2 - \omega_0^2}}e^{-\varepsilon t}\left(\varepsilon e^{\sqrt{\varepsilon^2 - \omega_0^2}t} + \sqrt{\varepsilon^2 - \omega_0^2}e^{\sqrt{\varepsilon^2 - \omega_0^2}t}\right.$$

$$\left. - \varepsilon e^{-\sqrt{\varepsilon^2 - \omega_0^2}t} + \sqrt{\varepsilon^2 - \omega_0^2}e^{-\sqrt{\varepsilon^2 - \omega_0^2}t}\right)$$

$$= x_{r0}e^{-\varepsilon t}\left(\frac{\varepsilon}{\sqrt{\varepsilon^2 - \omega_0^2}}\left(\frac{e^{\sqrt{\varepsilon^2 - \omega_0^2}t} - e^{-\sqrt{\varepsilon^2 - \omega_0^2}t}}{2}\right) + \right.$$

$$\left.\left(\frac{e^{\sqrt{\varepsilon^2 - \omega_0^2}t} + e^{-\sqrt{\varepsilon^2 - \omega_0^2}t}}{2}\right)\right) \tag{4.17}$$

Depending on the values of ε and ω_0 we have to distinguish three different cases:

4.1.1 Underdamped case ($\omega_0 > \varepsilon$)

In this case, $\sqrt{\varepsilon^2 - \omega_0^2}$ becomes imaginary, namely $j\sqrt{\omega_0^2 - \varepsilon^2}$. Equation (4.17) becomes:

$$x_r(t) = x_{r0}e^{-\varepsilon t}\left(\frac{\varepsilon}{\sqrt{\omega_0^2 - \varepsilon^2}}\left(\frac{e^{j\sqrt{\omega_0^2 - \varepsilon^2}t} - e^{-j\sqrt{\omega_0^2 - \varepsilon^2}t}}{2j}\right)\right.$$

$$\left. + \left(\frac{e^{j\sqrt{\omega_0^2 - \varepsilon^2}t} + e^{-j\sqrt{\omega_0^2 - \varepsilon^2}t}}{2}\right)\right) \tag{4.18}$$

With Euler's formulas: $\cos y = \dfrac{e^{jy} + e^{-jy}}{2}$ and $\sin y = \dfrac{e^{jy} - e^{-jy}}{2j}$ we can rewrite equation (4.18) to:

$$x_r(t) = x_{r0}e^{-\varepsilon t}\left(\frac{\varepsilon}{\sqrt{\omega_0^2 - \varepsilon^2}}\sin\left(\sqrt{\omega_0^2 - \varepsilon^2}t\right) + \cos\left(\sqrt{\omega_0^2 - \varepsilon^2}t\right)\right) \tag{4.19}$$

With the damping constant $h = \dfrac{\varepsilon}{\omega_0} = \sin\phi$ we obtain

$$\frac{\sin\phi}{\cos\phi} = \frac{\varepsilon/\omega_0}{\sqrt{1 - \varepsilon^2/\omega_0^2}} = \frac{\varepsilon}{\sqrt{\omega_0^2 - \varepsilon^2}} \tag{4.20}$$

and

$$x_r(t) = x_{r0}e^{-\varepsilon t}\left(\frac{\sin\phi}{\cos\phi}\sin\left(\sqrt{\omega_0^2 - \varepsilon^2}t\right) + \cos\left(\sqrt{\omega_0^2 - \varepsilon^2}t\right)\right)$$

$$= \frac{x_{r0}e^{-\varepsilon t}}{\cos\phi}\left(\sin\phi\sin\left(\sqrt{\omega_0^2 - \varepsilon^2}t\right) + \cos\phi\cos\left(\sqrt{\omega_0^2 - \varepsilon^2}t\right)\right) \tag{4.21}$$

With $\cos(\alpha - \beta) = \cos\alpha\cos\beta + \sin\alpha\sin\beta$ and $\omega = \sqrt{\omega_0^2 - \varepsilon^2}$ we get:

$$x_r(t) = \frac{x_{r0}}{\cos\phi} e^{-\varepsilon t} \cos\left(\sqrt{\omega_0^2 - \varepsilon^2}\, t - \phi\right)$$

$$= \frac{x_{r0}}{\cos\phi} e^{-\varepsilon t} \cos(\omega t - \phi)$$

$$\phi = \arcsin\left(\frac{\varepsilon}{\omega_0}\right)$$

$$(4.22)$$

In the underdamped case ($h < 1$), the seismometer oscillates with the period $T = \dfrac{2\pi}{\omega}$ which is always larger than the undamped natural period T_0:

$$T = \frac{2\pi}{\omega} = \frac{2\pi}{\sqrt{\omega_0^2 - \varepsilon^2}} = \frac{2\pi}{\omega_0\sqrt{1 - \varepsilon^2/\omega_0^2}} = \frac{2\pi}{\omega_0}\frac{1}{\sqrt{1 - h^2}}$$

$$= \frac{T_0}{\sqrt{1 - h^2}}$$

$$(4.23)$$

4.1.2 Overdamped case ($\omega_0 < \varepsilon$)

For the overdamped case, the damping constant h becomes greater than 1 and the solution of (4.13) becomes:

$$x_r(t) = A_1 e^{-\left(\varepsilon - \sqrt{\varepsilon^2 - \omega_0^2}\right)t} + A_2 e^{-\left(\varepsilon + \sqrt{\varepsilon^2 - \omega_0^2}\right)t}$$

$$= A_1 e^{-c_1 t} + A_2 e^{-c_2 t}$$

$$(4.24)$$

Since both c_1 and c_2 are real and positive, the solution will be an decaying exponential function. It will never oscillate.

4.1.3 Critically damped case ($\omega_0 = \varepsilon$)

For $\omega_0 \to \varepsilon \Rightarrow \sin(\sqrt{\omega_0^2 - \varepsilon^2}\,t) \to \sqrt{\omega_0^2 - \varepsilon^2}\,t$ and $\cos(\sqrt{\omega_0^2 - \varepsilon^2}\,t) \to 1$. Hence (4.19) returns to:

$$x_r(t) = x_{r0}(\varepsilon t + 1)\,e^{-\varepsilon t} \tag{4.25}$$

4.1.4 Comparison

In Fig. 4.2, the output signals for an initial displacement of -1 units corresponding to four different values of the damping constant are shown

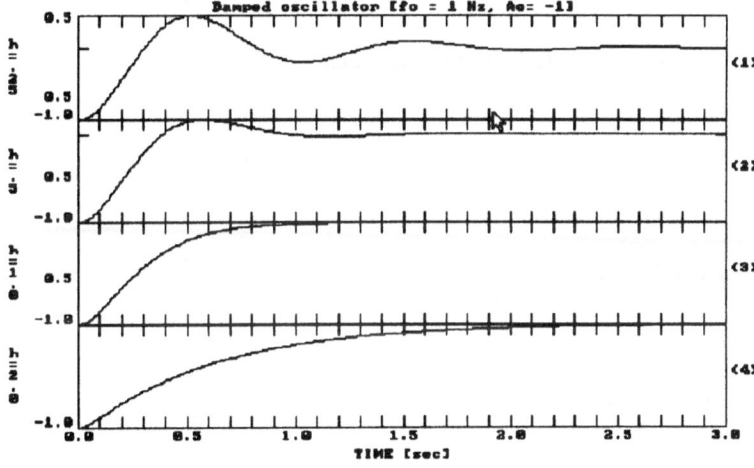

Fig. 4.2 Dependence of the output signal of a displacement seismometer ($f_0 = 1\,\text{Hz}$) on the damping constant h. From top to bottom h changes from .25, .5, 1.0, to 2.0. The initial displacement was assumed to be -1.

4.2 The determination of the damping constant

For the determination of the damping constant, let us rewrite (4.19)

$$x_r(t) = x_{r0}e^{-\varepsilon t}(c_1 \sin(\sqrt{\omega_0^2 - \varepsilon^2}t) + c_2 \cos(\sqrt{\omega_0^2 - \varepsilon^2}t)) \qquad (4.26)$$

Since the amplitude ratio of two consecutive maxima or minima are solely determined by the exponential term in equation (4.26), we can use this ratio to estimate the damping constant. We obtain:

$$\frac{a_k}{a_{k+2}} = \frac{e^{-\varepsilon t}}{e^{-\varepsilon(t+T)}} = \frac{e^{-\varepsilon t}}{e^{-\varepsilon t}e^{-\varepsilon T}} = e^{\varepsilon T} \qquad (4.27)$$

$$\ln\left(\frac{a_k}{a_{k+2}}\right) = \varepsilon T = \Lambda = \text{logarithmic decrement}$$

and

$$\frac{a_k}{a_{k+1}} = \frac{e^{-\varepsilon t}}{e^{-\varepsilon(t+T/2)}} = e^{\varepsilon(T/2)} \qquad (4.28)$$

$$\ln\left(\frac{a_k}{a_{k+1}}\right) = \frac{\varepsilon T}{2} = \frac{\Lambda}{2}$$

$$\Lambda = 2\ln\left(\frac{a_k}{a_{k+1}}\right) \qquad (4.29)$$

Between the logarithmic decrement Λ and the damping constant h, the following relationship exists:

$$\Lambda = \varepsilon T = \frac{\varepsilon T_0}{\sqrt{1-h^2}} = \frac{\varepsilon\dfrac{2\pi}{\omega_0}}{\sqrt{1-h^2}} = \frac{2\pi h}{\sqrt{1-h^2}} \qquad (4.30)$$

$$h = \frac{\Lambda}{\sqrt{4\pi^2 + \Lambda^2}}$$

(4.31)

Problem 4.1 Most seismometers operate on the principle of a moving coil within a magnetic field. Hence, they do not record the ground displacement but the ground velocity. Are equations (4.27) - (4.30) also valid for this kind of systems?

4.2.1 The determination of the damping constant for an electromagnetic sensor

Most commonly, the pendulum motion $x_r(t)$ (cf. Fig. 4.1) is measured by an electrodynamic sensor (Fig. 4.3)

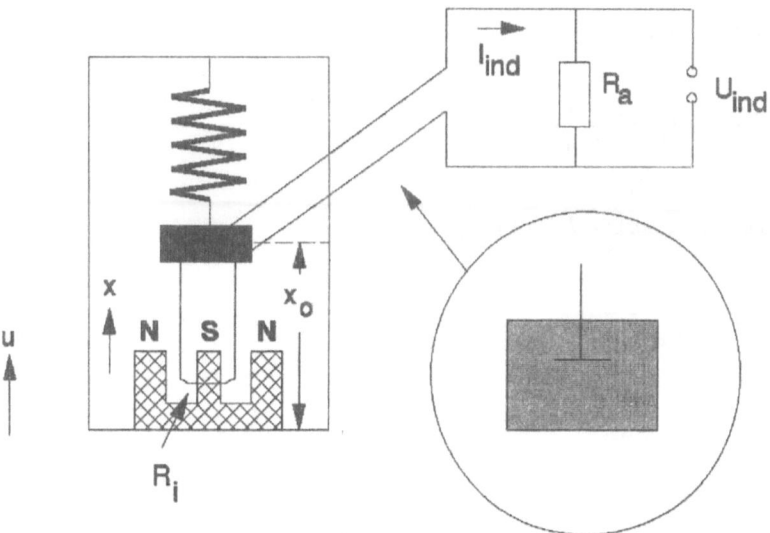

Fig. 4.3 Schematic model of an electromagnetic sensor. The dashpot in figure 4.1 is replaced by a coil moving in a magnetic field.

In this case, the dashpot in Fig. 4.1 is replaced by a coil which is fixed to the mass and moving through a permanent magnetic field. A voltage U_{ind} is generated across the coil proportional to $\dot{x}_r(t)$, the velocity of the seismometer mass with respect to the seismometer frame. If the coil is shunted by a resistance R_a, the generated current I_{ind} will be:

$$I_{ind} = \frac{U_{ind}}{R_a + R_i}$$ (4.32)

Here R_i is the internal resistance of the damping circuit including the coil. The corresponding magnetic field will be oriented in a way to damp the movement producing the voltage U_{ind}. The damping factor ε will be proportional to $\frac{1}{R_a + R_i}$.

$$\varepsilon \sim \frac{1}{R_a + R_i}$$ (4.33)

Taking into account the mechanical attenuation of the pendulum (ε_0) as well, we obtain ($h = \varepsilon / \omega_0$):

$$\varepsilon = h\omega_0 = \varepsilon_0 + b\frac{1}{R_a + R_i}$$ (4.34)

For the damping constant h, this becomes

$$h = h_0 + b'\frac{1}{R_a + R_i}$$ (4.35)

with h_0 describing the mechanical damping of the pendulum.

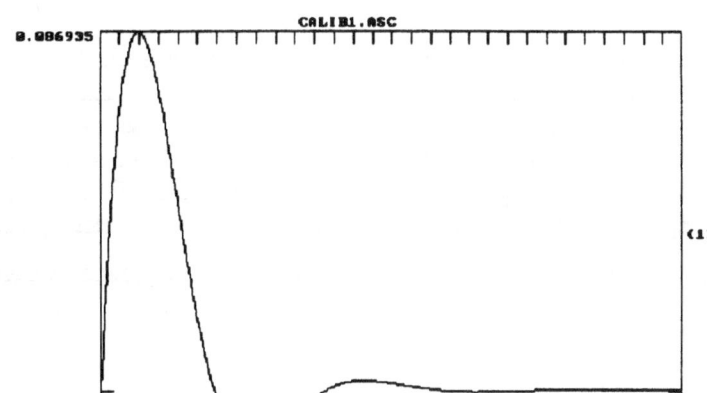

Fig. 4.4 Seismometer calibration pulse (response of an electrodynamic seismometer to a step function in acceleration)

In addition to the elements shown in Fig. 4.3, electromagnetic seismometers often have an additional calibration coil to impose a predefined displacement onto the seismometer mass. Switching the calibration current on/off produces a step function in acceleration which corresponds to releasing the seismometer mass from a constant initial displacement. An example for the response curve to such a calibration signal is shown in Fig. 4.4.

Problem 4.2 The calibration signal shown in Fig. 4.4 corresponds to the response of a velocity sensor to a step function in acceleration. a) What is the theoretical relationship between this signal and the displacement impulse response? b) The first few peak amplitude values are: 0.0869349, -0.014175, 0.00231063, -0.000376618, and 6.1422×10^{-5}. Determine the damping constant h and the natural frequency f_0 of the system.

4.3 The frequency response function

Looking at the solution of the seismometer equation (4.8) under the initial conditions described above provided us with a tool to obtain the system parameters. What we actually want to know, however, is the response of a seismometer to an arbitrary input signal. Since an arbitrary function can be described as a superposition of harmonics under very general conditions (Fourier series or Fourier integral), we are looking in some more detail at the solution of the seismometer equation to an harmonic input signal $u_g(t) = A_i e^{j\omega t}$. The corresponding ground acceleration is $\ddot{u}_g(t) = -\omega^2 A_i e^{j\omega t}$ and equation (4.8) becomes:

$$\ddot{x}_r(t) + 2\varepsilon \dot{x}_r(t) + \omega_0^2 x_r(t) = \omega^2 A_i e^{j\omega t} \tag{4.36}$$

Again, we are making the *Ansatz*:

$$x_r(t) = A_o e^{j\omega t}$$

$$\dot{x}_r(t) = j\omega A_o e^{j\omega t}$$

$$\ddot{x}_r(t) = -\omega^2 A_o e^{j\omega} \tag{4.37}$$

In general, A_i and A_o are complex quantities. Inserting (4.37) into (4.36), we get:

$$-\omega^2 A_o + 2\varepsilon j\omega A_o + \omega_0^2 A_o = \omega^2 A_i \tag{4.38}$$

and

$$\frac{A_o}{A_i} = \frac{\omega^2}{\omega_0^2 - \omega^2 + j2\varepsilon\omega} = U(j\omega) \tag{4.39}$$

$U(j\omega)$ is called the frequency response function. Separating the real and imaginary parts and after some manipulation we get:

$$\frac{A_o}{A_i} = \frac{\omega^2}{\sqrt{\left(\omega_0^2 - \omega^2\right)^2 + 4\varepsilon^2\omega^2}} e^{j\Phi} \tag{4.40}$$

with

$$\Phi = \Phi(\omega) = arc\tan\frac{-2\varepsilon\omega}{\omega_0^2 - \omega^2} = arc\tan\frac{-2h\omega_0\omega}{\omega_0^2 - \omega^2}$$

$$= arc\tan\frac{-2h\omega/\omega_0}{1 - (\omega/\omega_0)^2}$$

Rewriting equation (4.40) in terms of the damping constant $h = \varepsilon/\omega_0$ we get

$$|U(j\omega)| = \frac{\omega^2}{\sqrt{(\omega_0^2 - \omega^2)^2 + 4h^2\omega_0^2\omega^2}}$$

$$= \frac{\omega^2/\omega_0^2}{\sqrt{(1 - \omega^2/\omega_0^2)^2 + 4h^2\omega^2/\omega_0^2}}$$

$$= \frac{1}{\sqrt{(\omega^2/\omega_0^2 - 1)^2 + 4h^2\omega_0^2/\omega^2}} \tag{4.41}$$

with

$$\Phi(\omega) = arc\tan\frac{-2h\omega_0\omega}{\omega_0^2 - \omega^2} = arc\tan\frac{-2h\omega/\omega_0}{1 - \omega^2/\omega_0^2} \tag{4.42}$$

For an harmonic input signal with $\omega = \omega_0$, the phase shift is $\pi/2$ independent of the damping.

For an electrodynamic system, the output voltage is proportional to ground velocity (instead of displacement). In addition, it depends on the generator constant G of the seismometer coil. Equation (4.41) becomes

$$|U(j\omega)| = G\frac{\omega^3}{\sqrt{(\omega_0^2 - \omega^2)^2 + 4h^2\omega_0^2\omega^2}}$$

$$= G\frac{\omega^3/\omega_0^2}{\sqrt{(1 - \omega^2/\omega_0^2)^2 + 4h^2\omega^2/\omega_0^2}} \tag{4.43}$$

$$G = \frac{\text{output voltage}}{\text{ground velocity}} \quad \left[\frac{V}{m/\text{sec}}\right]$$

4.3.1 The transfer function

In order to obtain an expression for the transfer function, let us now solve the seismometer equation (4.8) using the Laplace transform. Assuming that for $t = 0$ all the initial conditions (x_r, $\dot{x}_r$, y_r, $\dot{y}_r = 0$, we obtain for the Laplace transform of equation (4-8) ($\ddot{x}_r(t) + 2\varepsilon\dot{x}_r(t) + \omega_0^2 x_r(t) = -\ddot{u}_g(t)$):

$$s^2 X_r(s) + 2\varepsilon s X_r(s) + \omega_0^2 X_r(s) = -s^2 U_g(s) \tag{4.44}$$

and

$$(s^2 + 2\varepsilon s + \omega_0^2) X_r(s) = -s^2 U_g(s) \tag{4.45}$$

For the transfer function we obtain:

$$T(s) = \frac{X_r(s)}{U_g(s)} = \frac{-s^2}{s^2 + 2\varepsilon s + \omega_0^2} \tag{4.46}$$

Since a quadratic equation $x^2 + bx + c = 0$ has the roots $x_{1,2} = -b/2 \pm \sqrt{b^2/4 - c}$, we get for the pole positions $p_{1,2}$:

$$p_{1,2} = -\varepsilon \pm \sqrt{\varepsilon^2 - \omega_0^2}$$

$$= -h\omega_0 \pm \omega_0 \sqrt{h^2 - 1}$$

$$= -(h \pm \sqrt{h^2 - 1})\,\omega_0 \qquad\qquad\qquad (4.47)$$

Problem 4.3 Calculate the impulse response (and spectrum) for a displacement seismometer with an eigenfrequency of 1 Hz and damping factors of $h = 0.25, 0.5, 0.62$, respectively. How do the locations of the poles change for the different damping constants?

Problem 4.4 How do we have to change the pole and zero distribution if we want to change the seismometer from Problem 4.3 into an electrodynamic system recording ground velocity?

5 The sampling process

In this chapter we take a first look at the effects of the sampling process. We will simulate the sampling of analog data and the reconstruction of analog data from sampled values (Whittaker reconstruction). We will find that an analog signal can only be reconstructed from its sampled values, if the frequency content of the signal to be sampled contains no energy at and above half of the sampling frequency (sampling theorem). We will simulate what happens if we deliberately violate this rule (aliasing effect).

5.1 The sampling of analog data

When we use a computer program to model continuous phenomena - like simulating the output voltage of a seismometer for certain boundary conditions - we normally do not think about the prerequisites we have to meet in order to do so. We just do it and assume the results are meaningful. In terms of system theory, however, we have performed an important transition. We have gone from a continuous system to a discrete system. That's fine as long as we are aware of the fact that there are some rules we better not violate.

The same transition from a continuous system to a discrete system takes place when we acquire data in digital form. There we are actually doing two different processes:

• *Sampling or discretization* — Taking discrete samples of a continuous data stream. The data could still be in analog representation after the sampling process.

• *Analog to digital conversion (quantization)* — For voltage signals, this stage is normally done using an electronic device which is called ADC, 'analog to digital converter'. After having gone through this process, the data are digital and discrete. We will treat this process in chapter 6 "Analog to digital conversion".

Below, some of the effects of discretization of 'continuous' signals are demonstrated using the *discretization tool* within PITSA. Of course, all the data traces in PITSA are already in digital form. We are only approximating a continuous signal by one which has been sampled at a sampling frequency which has to be much higher than the one at which we want to re-discretize the trace. Let's assume for now without proof that this is a valid approximation. After going through this chapter it should have become clear under which conditions we are allowed to do so. In addition, we need to reverse the process of discretization. In PITSA, a procedure called *Whittaker reconstruction* (Stearns, 1975) is used. Again, let's assume for now without proof that we can reverse the sampling process and get back the continuous signal.

In Fig. 5.1, the input signals for the simulation of the discretization process are displayed. They consist of three sinusoidal signals with frequencies of 3.5, 6.5, and 13.5 Hz.

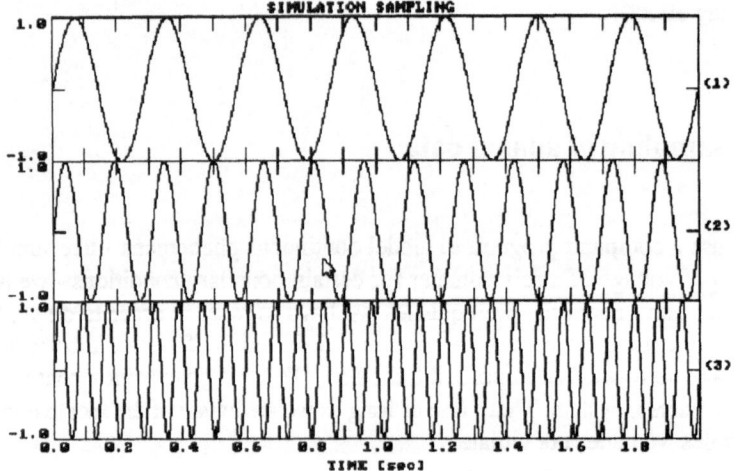

Fig. 5.1 Input signals for the simulation of the discretization process. The signal frequencies from top to bottom are 3.5, 6.5, 13.5 Hz, respectively.

During discretization, the 'continuous' input signals are 're-sampled' at the locations defined by the discretization frequency[1]. (Fig. 5.2).

---

1. When using the term discretization in the context of Fig. 5.2, what is actually done is zeroing out all but the 'discretized' samples without changing the internal sampling rate. An input sequence of n points will still contain n points after discretization in PITSA. This is different when using the option resampling or A/D conversion.

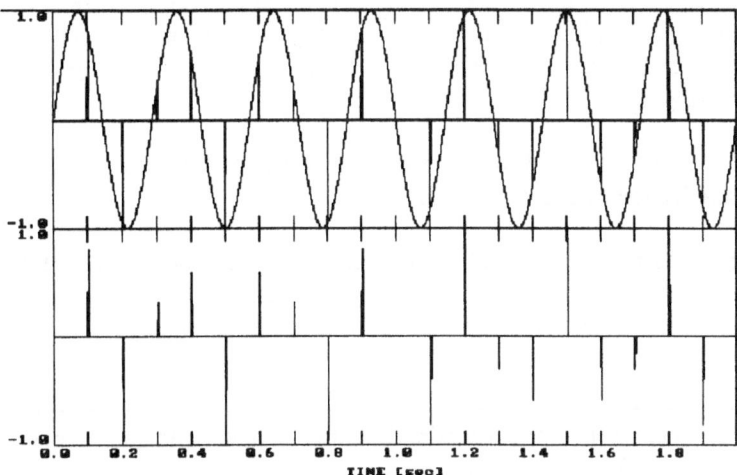

Fig. 5.2 Discretizing channel 1 of Fig. 5.1 using a discretization frequency of 10 Hz. The vertical bars show the locations and the values of the samples.

Fig. 5.3 shows the result of discretizing all the traces in Fig. 5.1 using a discretization frequency of 10 Hz and reconstructing them again into a 'continuous' representation using a Whittaker reconstruction.

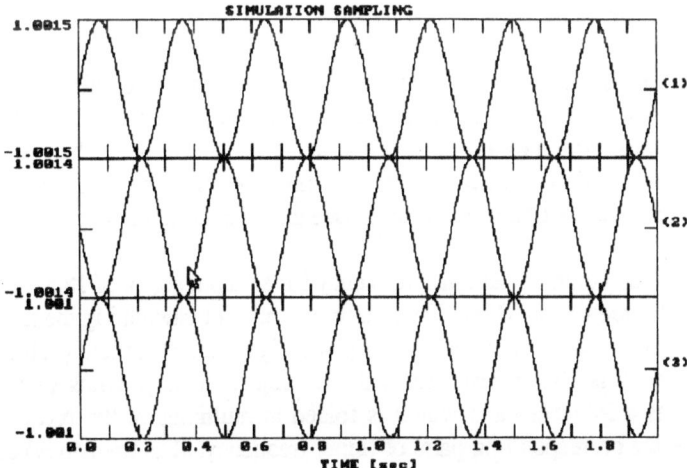

Fig. 5.3 Reconstructed traces 1-3 of Fig. 5.1 (after discretizing all of them with 10 Hz prior to reconstruction). Notice that channel 2 is just phase shifted by π in comparison to channel 1 and 3.

You can see from comparison Fig. 5.3 and Fig. 5.1 that only trace 1 is reconstructed correctly. Reconstructed trace 2 looks like a phase shifted version of trace 1. The reconstructed trace 3 is identical to trace 1. Although the original signals had quite different frequencies, the frequencies of the reconstructed traces in this special case are identical (Fig. 5.3).

Problem 5.1 What is the highest frequency which can be reconstructed correctly using a discretization frequency of 10 Hz?

Once a signal has been discretized at a sampling frequency too low for the actual frequency content, the signal can not uniquely be reconstructed. In order to reconstruct a continuous signal from its sampled values, the following rule has to be obeyed:

• *Sampling theorem* — For a continuous time signal to be uniquely represented by samples taken at a sampling frequency of f_{dig}, (every $1/f_{dig}$ time interval), no energy must be present in the signal at and above the frequency $1/2\ f_{dig}$. $1/2\ f_{dig}$ is called the Nyquist frequency. Signal components with energy above the Nyquist frequency will be mapped by the sampling process onto the so called alias frequencies within the frequency band of 0 to Nyquist frequency. This effect is called the *alias* effect.

An example for the alias effect can be seen in old western movies. If the wheels of a stagecoach seem to turn backwards, then the sampling of the images was to slow to catch the movement of the wheels uniquely.

In order to meet the requirements of the sampling theorem, analog signals have to be filtered in a way to remove all the unwanted frequency components (Anti Alias Filter). The reconstruction of a continuous signal from its sampled values (e. g. the Whittaker reconstruction) is also done by low pass filtering. The Whittaker reconstruction simply takes out all frequencies above the Nyquist frequency.

Problem 5.2 What would be the alias frequency for an input signal of 18.5 Hz and a discretization frequency of 10 Hz? Try to infer the rule for calculating the alias frequency for a given signal frequency and a given digitization frequency. Hint: The Nyquist frequency is also referred to as the folding frequency. Think of the frequency band as a foldable band which is folded at multiples of the Nyquist frequency. Mark the corresponding pairs of (input frequency, alias frequency) on this band. For problem 5.1, this would be (3.5, 3.5), (6.5, 3.5) (13.5, 3.5). If you have a lack of imagination, it may actually help to cut out a paper band and to actually perform the folding.

6 Analog to digital conversion

In this chapter we will look at the properties of analog to digital conversion (ADC) and the limitations they introduce to sampled data. We will discuss the relationship between resolution and dynamic range of A/D converters (ADCs). We will end this chapter by simulating techniques for improving the dynamic range of ADCs by gain-ranging and oversampling.

6.1 The principle of analog to digital conversion

We have seen that the timing of the sampling process puts limits onto the frequency contents of the data to be sampled. The way we were simulating the sampling process in PITSA, was that we were taking a subset of the original input sequence. The important point is, that the amplitudes for this subset were identical to the original values. When we are doing a conversion of a continuous analog input signal into a digital sequence of samples using electronic analog to digital converters on voltage signals, this is no longer true.

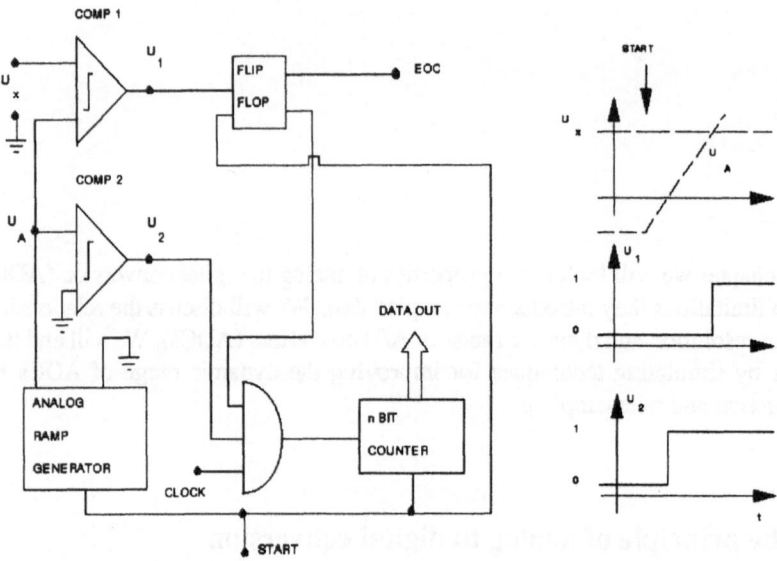

Fig. 6.1 Principle of a single slope analog to digital converter (after Jaeger, 1982).

To understand the sources of errors involved, we are going to look at the working principles of a simple type of analog to digital converter, the single slope ADC (Fig. 6.1). It is assumed that an input voltage U_x is going to be converted into a discrete, digital value. The voltage U_x is assumed to be constant during the time needed for the conversion. For this example we also assume (without loss of generality) U_x to be positive. A start signal initiates the actual conversion and starts an analog ramp generator which produces a voltage U_A which linearly increases with time (cf. Fig. 6.1 uppermost right panel). The voltage U_A is checked by a comparator (COMP 2) to determine whether it is equal or larger to 0 voltage. Once it reaches 0 voltage, output line U_2 of COMP 2, which is connected to a logic element, goes high. The line between the logic element (AND element) and the *n bit counter* will go high if all the input lines are high. In addition to the comparator 2, the remaining input lines of the logic element are connected to a clock (which goes high every clock cycle) and a flip flop. A flip flop is a bistable oscillator which can only have two different states (HIGH or LOW). Since the flip flop is assumed to start out in a high state, once the voltage U_2 is high, the counter will receive a voltage pulse every clock cycle. All the counter does is counting the pulses it receives. The counter, however, has only a certain number of bits available to store the result of its counting process. For n bits, it can only count from 0 to $2^n - 1$.

Once the counting process has started, the voltage U_A is increasing. At a second comparator (COMP 1), it is continuously compared to the input voltage U_x. Once $U_A \geq U_x$, a high output voltage U_1 will toggle the flip flop off. This in turn will prohibit the logic element to send a signal to the counter. In other words, the conversion stops.

The input voltage U_x is converted into a digital value by counting the time it takes for the ramp generator to produce a voltage as high as the input voltage. If the counter has n bits for storing the result of its counting, the ADC is called a n-bit ADC, which means it has 2^n output states. Input voltages ranging from 0 to the full scale range voltage of the ADC are mapped onto discrete values between values from 0 to $2^n - 1$. For a 3 bit ADC and a full scale input voltage of 10 V, this situation is shown in Fig. 6.2.

The smallest input voltage change that can cause a change of the output value of the ADC, is called quantum or least significant bit (LSB). For a n-bit ADC, it is given by

$$Q = 1\text{LSB} = \frac{\text{FULL SCALE VOLTAGE}}{2^n} \tag{6.1}$$

Since the value of the quantum which controls the resolution of the ADC is directly depending on the number of bits of the ADC, a n-bit ADC is said to have n bits of *resolution*. For the linear increasing input signal in Fig. 6.2, the probability density function of the error signal is flat between $\pm Q/2$ with a variance of $Q^2/12$ (Oppenheim and Schafer, 1989, p. 120). As can be imagined from Fig. 6.2, the error of the ADC can strongly depend on the input signal.

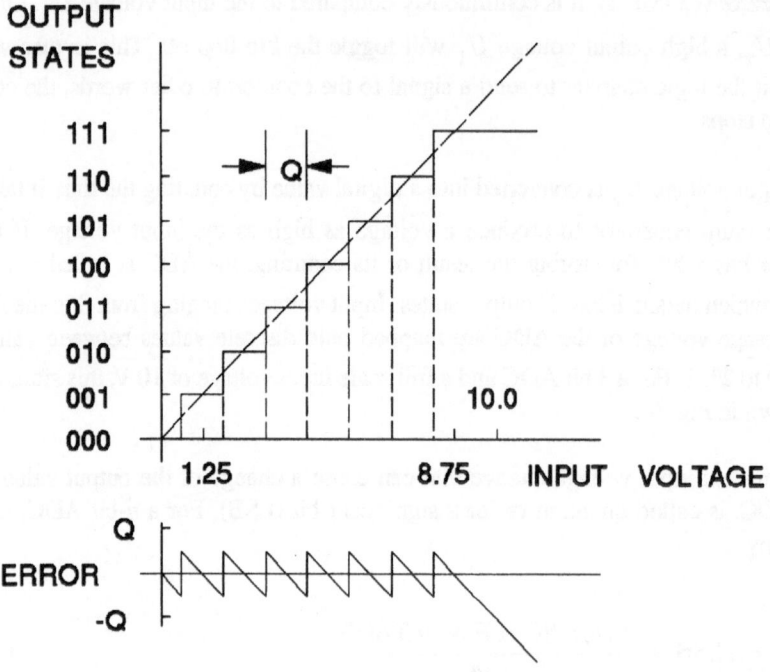

Fig. 6.2 Mapping of input voltage to output states for a 3 bit ADC. The lower panel shows the error signal. Q corresponds to the quantum or least significant bit (LSB).

An additional measure of the quality of an ADC is the *dynamic range*. For analog signals, the dynamic range is defined as the ratio between the largest and the smallest signal which can be measured. It is expressed in decibels (dB).

$$D = 20\log_{10}(A_{max}/A_{min}) \qquad [dB] \qquad (6.2)$$

For a n bit ADC, the dynamic range is defined as:

$$D = 20\log_{10}(2^{n}) \qquad [dB] \qquad (6.3)$$

1 bit = 6.021 dB

The ADC in Fig. 6.2 would be said to have a resolution of 3 bits and a dynamic range of 18 dB. In Fig. 6.3 it is demonstrated what happens if an ADC is fed with an input signal with larger amplitudes than it can represent. The upper trace in Fig. 6.2 shows a decaying sinusoid with a maximum amplitude of 839.46. Below the output signal for an ADC with a resolution of 12 bit and a LSB of 0.1 is shown.

Notice   that   the   output   trace   is   saturated   at   an   amplitude   of
$0.1 \cdot (2^{12} - 1) = 0.1 \cdot 2047 = 204.7$.

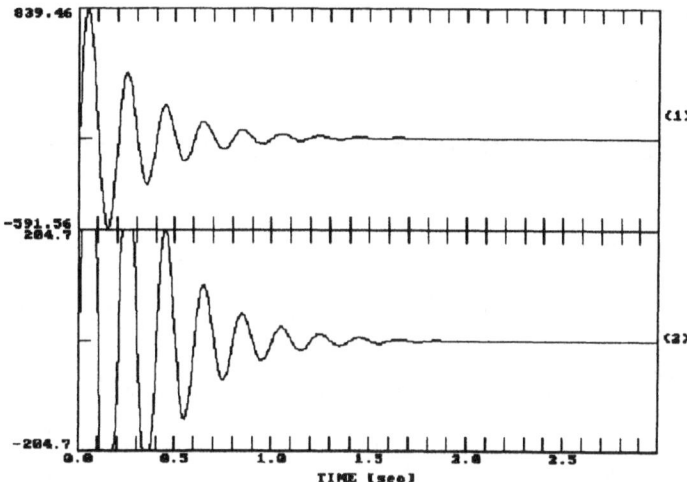

Fig. 6.3 The saturation effect for an ADC with insufficient dynamic range.

A simple way to get rid of the saturation (clipping) problem would be to attenuate
the input channel before it is fed into the ADC as shown in Fig. 6.4.

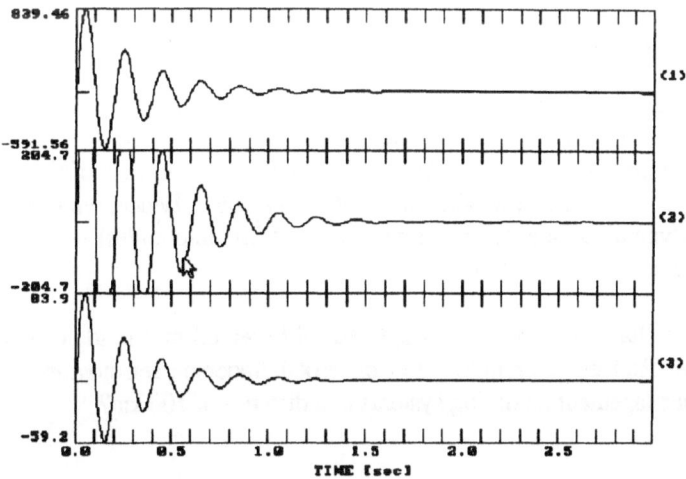

Fig. 6.4 Removing the saturation effect by attenuating the input signal using an attenuation
by a factor of 10 (cf. channel 3 and 2).

In Fig. 6.4, the ADC input signal (top trace) is attenuated by a factor of 10. before being fed into the ADC with the same parameters as being used for Fig. 6.3. As can be seen in the lowermost trace, the dynamic range is now sufficient to convert the complete waveform without saturation. However, what we run into using this simple minded approach is a different problem which can be seen if we are looking at the later portion of the resampled signal (Fig. 6.5).

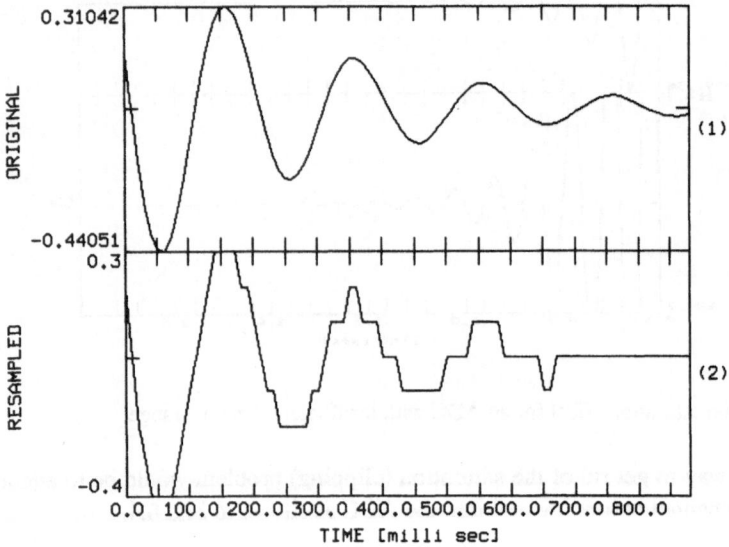

Fig. 6.5 Lack of resolution.

The original trace after attenuation is shown in the top trace of Fig. 6.5 while the bottom trace shows the result of applying an ADC to the top trace using an LSB of 0.1 and a resolution of 12 bit. For the displayed data window, most of the changes in signal amplitude are below what the ADC can resolve. Hence, we have traded the gain in dynamic range (which helped with the big amplitudes) with a lack of 'resolution' for the small amplitudes.

Problem 6.1 What is the dynamic range needed to record magnitude 0 as well as magnitude 6 (Richter magnitude based on Wood-Anderson seismographs) earthquakes on displacement recording systems in a distance of 100 km?

6.2 Increasing dynamic range and resolution

Since the dynamic range needed for on scale recording of a reasonable magnitude range of interest is still touching the limits of currently available ADCs at reasonable costs, people have found ways to increase both the dynamic range as well as the resolution. Two different methods are going to be discussed and simulated within this context. *Gain ranging* and *oversampling*. The first technique trades dynamic range for resolution while the second technique increases the resolution by decreasing the influence of quantization noise and consequently increases the dynamic range.

6.2.1 Gain ranging

Fig. 6.6 sketches the structure of a gain ranging ADC. The key elements are a regular n bit ADC, a programmable gain amplifier (PGA), and a control logic. As long as the analog input signal is small in comparison with the full scale range of the ADC, the gain ranging ADC works just like a plain ADC. However, once the input signal reaches a certain level ('switch up fraction', e.g $0.25 \cdot$ of the FULL SCALE RANGE of the ADC) the control logic will cause the PGA to decrease the gain of the pre-amplifier. As a consequence, the signal will again fit into the full scale range of the ADC easily. If the signal amplitude is further increasing, the gain of the pre-amplifier will be switched down again and again.

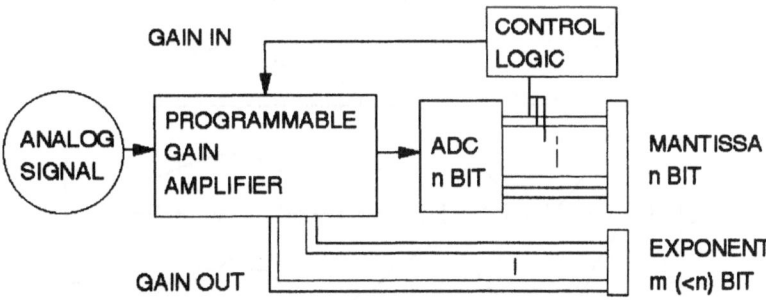

Fig. 6.6 The principle of a gain ranging ADC.

On the other hand, if the signal amplitude has been below a certain threshold for a certain number of cycles, the control logic will cause the PGA to switch up the gain again.

What is recorded in a gain ranging ADC, are both the output values of the plain

ADC (mantissa) and the status of the PGA (gain). If we use m bits for recording the
gain status, we can record 2^m gain states. Since the PGA is switching gain in powers of 2, the actual gain changes corresponding to 2^m gain states makes up a factor
of 2^{2^m}. The dynamic range of a gain ranging ADC of m + n bits is therefore:

$$D_{gr} = 20\log_{10}(2^n \cdot 2^{2^m}) \quad [\text{dB}]$$

$$= 20\log_{10}(2^{n+2^m}) \quad [\text{dB}]$$

(6.4)

For a given number of total bits, the increase in dynamic range is traded for a gain
dependent change of resolution. In Fig. 6.7, the data trace in Fig. 6.3 is resampled
using a gain ranging ADC with 8 bit of resolution for the mantissa and 4 bits for the
exponent (negative gain). From the 16 (2^4)possible states of the programmable
gain amplifier (PGA), however, only 11 are used in this example (in real gain ranging ADCs, some bits of the exponent are sometimes used as error flag). Visually we
do not see a difference between the result of the gain ranging ADC (bottom trace)
and the original input trace (top trace) even in the 'low amplitude' range of the signal. For comparison, the second and the third trace in Fig. 6.7 show the corresponding data window for trace 2 in Fig. 6.3 (insufficient dynamic range) and trace 2 in
Fig. 6.5 (insufficient resolution), respectively. We can see that the result of the gain
ranging ADC is visually indistinguishable from the input trace.

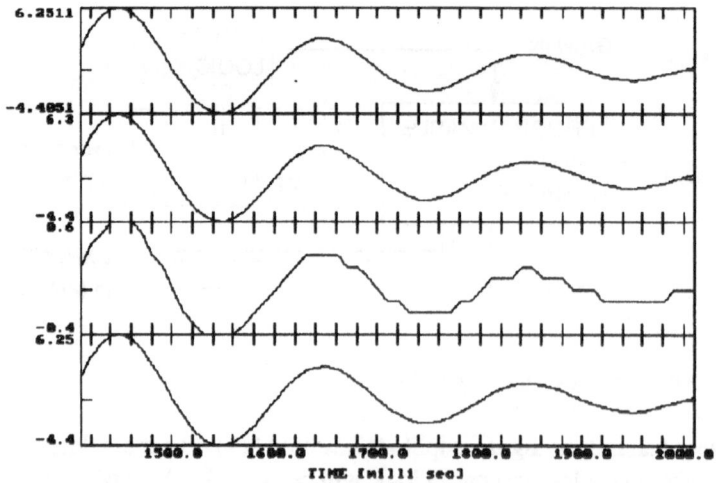

Fig. 6.7 Resolution and dynamic range of plain- and gain ranging ADC.

As final example for the action of gain ranging ADCs, Fig. 6.8 shows from top to bottom the input trace, the quantized values, the negative gain, and the error signal for a half wave of a cosine signal with a maximum amplitude of 2000. The resolution was chosen to be 12 bit, the LSB = 0.1, and the number of possible gain states as 11.

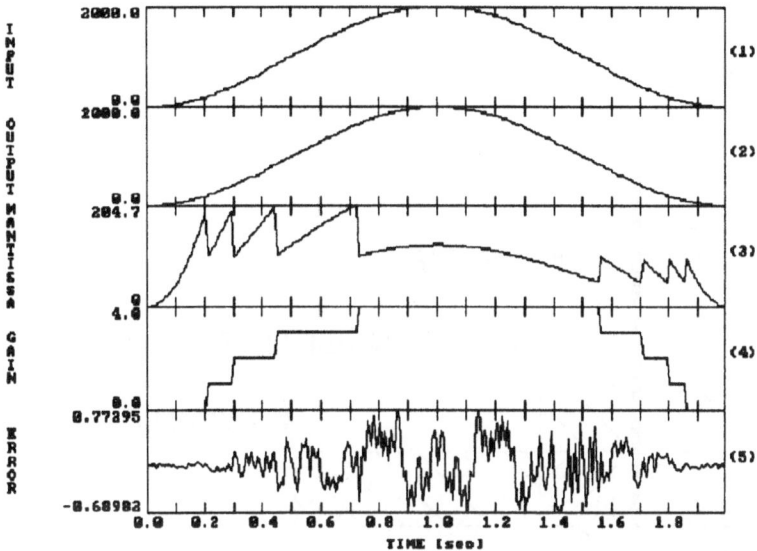

Fig. 6.8 Input signal, quantized value, mantissa, negative gain states, and error signal for a gain ranging ADC[1].

The quantized value in channel 2 equals (channel 3) x $2^{(\text{channel } 4)}$. The error signal in channel 5 is the difference between the input signal and the quantized values. As a consequence of the tradeoff between dynamic range and resolution for gain ranging ADCs, the error signal becomes dependent on the gain state.

6.2.2 Oversampling

A different technique which has become popular in the context of seismic recording only recently is based on the assumption that the variance of the quantization noise is independent of the sampling rate under certain conditions (Oppenheim and

---

1. In the present example, the logic for the gain switching is different depending on whether the signal is increasing or decreasing its amplitude. This is also the case in some real ADCs.

Schafer, 1989). If the quantization noise has a probability density function which is flat between $\pm Q/2$ (cf. Fig. 6.2), the variance of the quantization noise is $Q^2/12$ (Q = quantum of the ADC) which is independent of the sampling rate. Hence, following Parseval's theorem, the area covered by the quantization noise in the frequency domain is independent of the sampling rate. This situation is sketched in Fig. 6.9

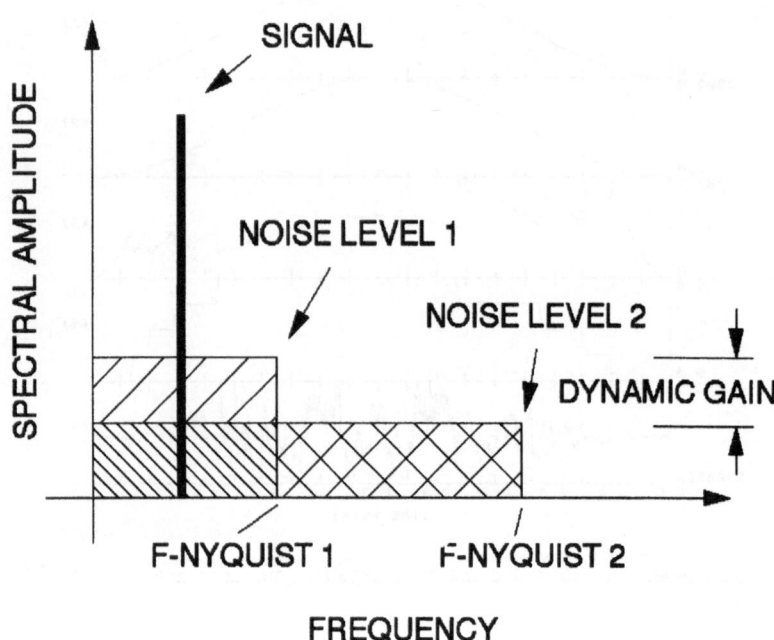

Fig. 6.9 Noise reduction by oversampling.

In order for the noise areas (hatched areas) corresponding to the two given Nyquist frequencies (F-NYQUIST 1 and F-NYQUIST 2, in Fig. 6.9, respectively) to be equal, $\frac{\text{NOISE LEVEL 1}}{\text{NOISE LEVEL 2}}$ has to be equal to $\frac{\text{F-NYQUIST 2}}{\text{F-NYQUIST 1}}$. By sampling the input signal at a much higher frequency than finally desired (corresponding to F-NYQUIST 2), and subsequently low pass filtering and digitally resampling without further loss of resolution, the available dynamic range (signal to noise ratio) will be larger than the case in which the input signal is sampled at the lower sampling frequency directly.

Fig. 6.10 shows two synthetic signals which will be used to demonstrate the effects exploited in this context. The top trace shows a noisefree sinusoid with a maximum amplitude of 0.1 and a signal frequency 10 Hz, created at a digitization frequency of 1000 Hz. The bottom trace contains the same signal, however, superimposed by Gaussian noise with a variance of 0.5.

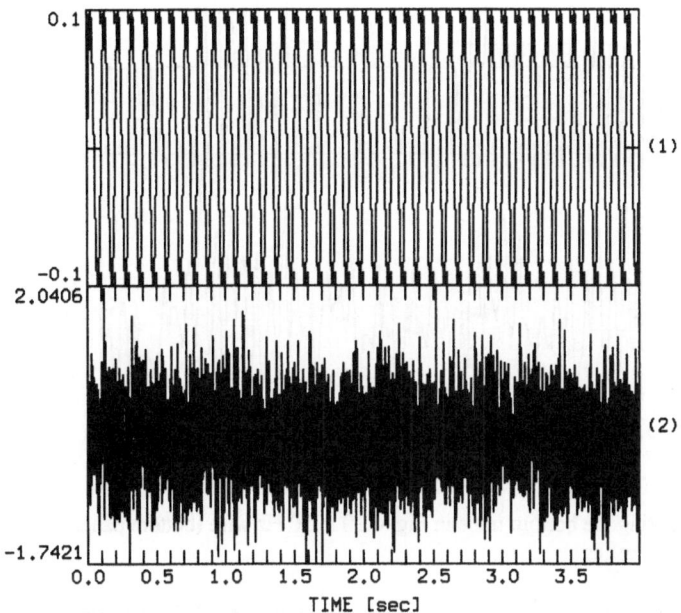

Fig. 6.10 Synthetic traces to demonstrate the effects of oversampling. The top trace shows a sinusoid of 10 Hz signal frequency and a peak amplitude of 0.1. The bottom trace shows the same signal superimposed by Gaussian noise with a variance of 0.5.

We will use two different approaches to decimate the bottom trace down to 50 Hz using 12 bits of resolution and a LSB of 0.2. Notice, that the LSB of 0.2 is larger than the peak amplitude of the signal. Hence, sampling the top trace using these values would result in a zero trace. Nevertheless, we will see that we will be able to resolve the 10 Hz signal to a reasonable degree in the presence of noise. The presence of noise also assures that the quantization noise is sufficiently well-distributed in the frequency domain (cf. Fig. 6.9) for the oversampling technique to work.

Fig. 6.11 shows the result of decimating the bottom trace in Fig. 6.10 to 50 Hz. The top trace is the result of antialias filtering at 15 Hz, followed by resampling at 50 Hz using an LSB of 0.2 and 12 bits of resolution. The bottom trace was resampled at 1000 Hz (corresponds to oversampling) with the same LSB and resolution, subsequently antialias filtered and digitally resampled (at full floating point resolution) to 50 Hz. For the digital antialias filter, a 4 pole, zero-phase Butterworth filter was used in both cases.

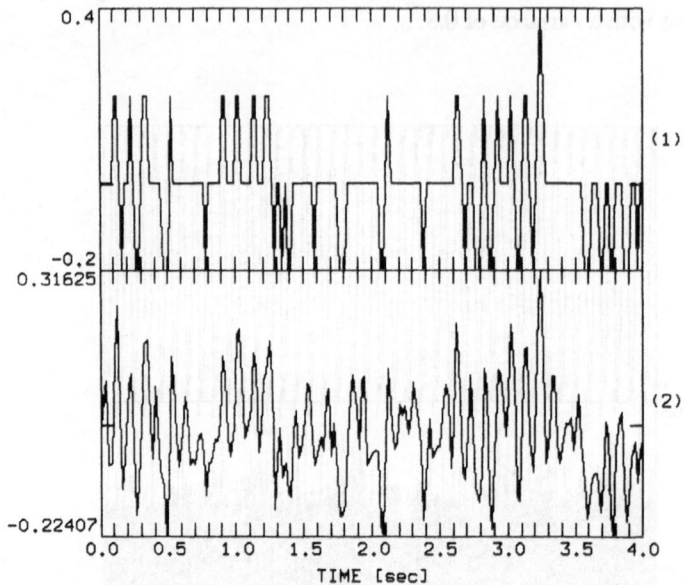

Fig. 6.11 Decimating the bottom trace in Fig. 6.10 to 50 Hz with (bottom panel) and without (top panel) oversampling.

The spectra of the decimated traces are shown in Fig. 6.12. Notice that except for the different ways to obtain the final sampling rate and resolution, both traces have been treated identically within the frequency band displayed.

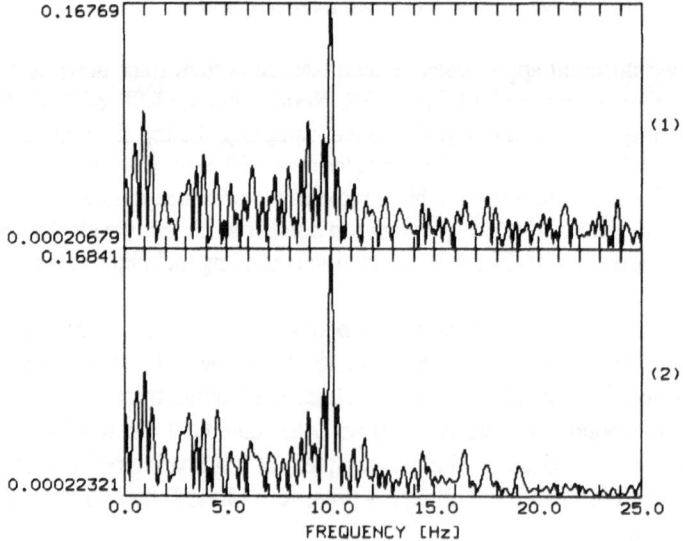

Fig. 6.12 Amplitude spectra of traces in Fig. 6.11.

Both Fig. 6.11 and Fig. 6.12 show that the quantization error is less, if the trace is originally sampled at higher frequency and subsequently decimated digitally at full floating point resolution. This technique has become one of the standard methods employed in modern digital recording systems. Digital seismograms from these kind of instruments can be seen as the output of a sequence of filter stages which commonly involve both *analog* and *digital* components (Fig. 6.13).

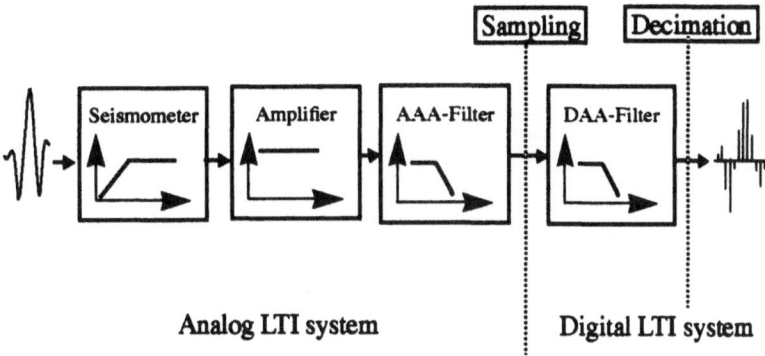

Fig. 6.13 Building blocks of modern seismic aquisition systems. AAA-filter denotes the analog antialias filter, whereas DAA-filter is the digital antialias filter to be applied before decimation.

In addition to the gain in dynamic range, this allows instrument designers to use fairly simple analog antialias filters and do most of the work in the digital domain where powerful and easy to implement digital antialias filters can be used.

Both Figs. 6.11 and Fig. 6.12 show that the quantization error is less if the data is originally sampled at higher frequency and subsequently decimated digitally at full floating-point resolution. This technique has become one of the standard methods employed in modern digital recording systems. Digital seismograms from these kind of instruments can be seen as the output of a sequence of filter stages which commonly involve both analog and digital components (Fig. 6.13).

Fig. 6.13 ... Analog FIR system ... Digital FIR system ...

In addition to the gain in dynamic range, this allows inexpensive circuitry to be used.

7 From infinitely continuous to finite discrete

Up to this point, the concepts we have been using to describe the properties of LTI systems were developed for continuous LTI systems, corresponding to the analog part in Fig. 6.13. On the other hand, when we have been using PITSA to demonstrate these properties such as for calculating the frequency response function for a given pole-zero distribution, we were of course working with sequences of numbers. This was acceptable in the context of the examples we have been dealing with so far, however, in a general context we need to be well aware of some important differences between discrete and continuous systems. By now we should have enough intuitive feeling about some of the most important system properties so that we can extend our view by a more formal approach. This not only will provide us with additional tools for actual data processing, but also will give us some insight into the links between the worlds of the *infinitely continuous* and the *finite discrete*. In the following the notation of Oppenheim and Schafer (1989) will be used.

7.1 Fourier transform of continuous-time signals

Let us start out with a review of some of the elementary properties of the Fourier transform for continuous-time signals. In the context in which we have been using it, the Fourier transform can be thought of as mapping a 'time' signal $x(t)$ from the 'time domain' onto a 'frequency' signal $X(f)$ in the 'frequency domain' by the following equation:

$$\mathcal{F}\{x(t)\} = X(f) = \int_{-\infty}^{\infty} x(t)\, e^{-j2\pi ft}\, dt \tag{7.1}$$

Of course, the meaning of 'time-' and 'frequency domains' are not restricted to physical time and frequency. It is common practice to write the frequency in terms of angular frequency $\omega = 2\pi f$, and since (7.1) describes a complex quantity to write:

$$X(j\omega) = \int_{-\infty}^{\infty} x(t) e^{-j\omega t} \, dt \tag{7.2}$$

For the back transformation we have:

$$x(t) = \frac{1}{2\pi} \int_{-\infty}^{\infty} X(j\omega) e^{j\omega t} \, d\omega \tag{7.3}$$

One reason for using $j\omega$ is that this allows us right away to see the Fourier transform as a special case of the Laplace transform evaluated on the imaginary axis $(j\omega)$ of the complex s plane (cf. chapter "The impulse response" on page 19). The Fourier transform is a linear transform, which means that the transform of the scaled sum of two signals is the scaled sum of the individual transforms. In the following, we give a list of most important transform properties ($\Leftrightarrow$ indicates a transform pair, $x(t) \Leftrightarrow X(j\omega)$):

- *Time shifting* — $x(t-a) \Leftrightarrow X(j\omega) e^{-j\omega a}$ for $a > 0$ (7.4)

- *Derivative* — $\dfrac{d}{dt} x(t) \Leftrightarrow j\omega X(j\omega)$ (7.5)

- *Integration* — $\displaystyle\int_{-\infty}^{t} x(\tau) \, d\tau \Leftrightarrow \dfrac{1}{j\omega} X(j\omega)$ (7.6)

In case the signal has certain symmetry properties in one domain such as being even or odd, etc. it will have corresponding symmetry properties in the other domain (Tab. 7.1).

Table 7.1 Symmetry properties of the Fourier transform

time domain	frequency domain
$x(t)$ real	$X(-j\omega) = [X(j\omega)]^*$
$x(t)$ imaginary	$X(-j\omega) = -[X(j\omega)]^*$
$x(t)$ even	$X(-j\omega) = X(j\omega)$
$x(t)$ odd	$X(-j\omega) = -X(j\omega)$
$x(t)$ real and even	$X(j\omega)$ real and even
$x(t)$ real and odd	$X(j\omega)$ imaginary and odd
$x(t)$ imaginary and even	$X(j\omega)$ imaginary and even
$x(t)$ imaginary and odd	$X(j\omega)$ real and odd
	* denotes complex conjugate

These symmetry properties are used for once to increase the computational efficiency for the calculation of Fourier transforms but they are also helpful to understand some properties of linear filters.

Of special importance in the context of filtering are the properties of the Fourier transform with respect to the convolution of two signals. The convolution of two functions h(t) and g(t) is defined as

$$g * h = \int_{-\infty}^{\infty} g(\tau) h(t-\tau) d\tau \qquad (7.7)$$

If the corresponding spectra are denoted $H(j\omega)$ and $G(j\omega)$, the transform pair is given by the

• *Convolution Theorem* — $g(t) * h(t) \Leftrightarrow G(j\omega) H(j\omega)$

In earlier chapters, we have already made use of this theorem without explicitly stating it. It becomes extremely important in the context of deconvolution, since it tells us that we can remove the effect of a filter by simply dividing the spectrum of the signal by the frequency response function of the system. In the time domain this operation has no equally simple equivalence.

7.2 Fourier transform of discrete-time signals

When dealing with discrete-time signals, the Fourier transform is defined as

$$\mathcal{F}\{x[n]\} = X(e^{j\omega}) = \sum_{n=-\infty}^{\infty} x[n] e^{-j\omega n} \tag{7.8}$$

Here, $x[n]$ denotes a discrete-time sequence which may be defined for infinitely many integer values (a function whose domain is integer) and ω is the angular frequency. Together with the inverse transform:

$$\mathcal{F}^{-1}\{X(e^{j\omega})\} = x[n] = \frac{1}{2\pi} \int_{-\pi}^{\pi} X(e^{j\omega}) e^{j\omega n} d\omega \tag{7.9}$$

equation (7.8) and (7.9) form a Fourier transform pair (Oppenheim and Schafer, 1989). The Fourier transform for a discrete-time signal is always a periodic function with a period of 2π. This can easily be seen by replacing ω in (7.8) by $\omega + 2\pi$ which results in $X(e^{j(\omega + 2\pi)}) = X(e^{j\omega})$ or more general $X(e^{j(\omega + 2\pi r)}) = X(e^{j\omega})$ for any integer r. Since for the synthesis of $x[n]$ frequencies of ω and $\omega + 2\pi$ can not be distinguished, only a frequency range of length 2π needs to be considered. Since the frequency response function of a linear system equals the Fourier transform of the impulse response, *the frequency response function of a discrete-time system is always a periodic function.*

It is interesting to note at this point that the argument of the integral to obtain the Fourier series representation of a periodic continuous-time signal is of the same form as (7.8). Note also that in contrast to the continuous-time case, there is an asymmetry between equations (7.8) and (7.9). This corresponds to a modification of the equivalence between convolution and multiplication. While convolution of sequences corresponds to multiplication of periodic Fourier transforms, multiplication of sequences corresponds to periodic convolution (with the convolution only carried out over one period) of the corresponding Fourier transforms.

In the case of digital seismic signals, we are dealing with discrete-time signals which are obtained by taking samples from continuous signals at discrete times nT with T being the sampling interval in sec. In order to see the consequences of sampling in the frequency domain more formally, we can view this process as being caused by the multiplication of the continuous signal with a periodic impulse train

$$s(t) = \sum_{n=-\infty}^{\infty} \delta(t-nT) \tag{7.10}$$

with $\delta(t-nT)$ being time shifted delta functions. If the continuous-time signal would be $x_c(t)$, the sampled signal can be written

$$x_s(t) = x_c(t)\, s(t) \tag{7.11}$$

The Fourier transform of $x_s(t)$ is the convolution of the Fourier transforms of $s(t)$ and $x_c(t)$ which are denoted $S(j\omega)$ and $X_c(j\omega)$, respectively. From the properties of the delta function and the shifting theorem it follows that

$$S(j\omega) = \frac{2\pi}{T} \sum_{k=-\infty}^{\infty} \delta(\omega-k\omega_s) \tag{7.12}$$

with ω_s being the sampling frequency in rad/sec. Convolving $S(j\omega)$ and $X_c(j\omega)$ yields

$$X_s(j\omega) = \frac{1}{T} \sum_{k=-\infty}^{\infty} X_c(j\omega-kj\omega_s) \tag{7.13}$$

Hence the Fourier transform of a sampled signal is periodic with sampling frequency ω_s. If sampling has been done with proper anti-aliasing filtering, the Fourier transform of the continuous signal $X_c(j\omega)$ can be obtained from one period of the Fourier transform of the sampled signal by

$$X_c(j\omega) = T \cdot X_s(j\omega) \tag{7.14}$$

7.3 The z-transform

While the analog part of the recording system shown in Fig. 7.2 can be described in terms of the Laplace transform of its impulse response (transfer function $T(s)$), the digital part is commonly described using its discrete counterpart, the z-transform of the discrete impulse response which again is called transfer function $T(z)$. The bilateral z-transform of a discrete sequence $x[n]$ is defined as

$$Z\{x[n]\} = \sum_{n=-\infty}^{\infty} x[n]z^{-n} = X(z) \tag{7.15}$$

The z-transform transforms the sequence $x[n]$ into a function $X(z)$ with z being a continuous complex variable. We can visualize how the complex s plane is mapped onto the z plane if we substitute $z = e^{sT} = e^{\sigma T}e^{j\omega T} = re^{j\omega T}$. This is sketched in Fig. 7.1. For points on the imaginary axis ($s = j\omega$) r becomes 1 and z becomes the unit circle. In this case, equation (7.15) reduces to the Fourier transform of $x[n]$ (see equation (7.8)). This again underlines the statement about the z-transform being the discrete counterpart of the Laplace transform. The origin ($\omega = 0$) maps onto $z = 1$ while $j\omega = j\omega_s/2 = j\pi/T$ maps onto $e^{j\pi} = \cos(\pi) = -1$. Positive frequencies are mapped onto the upper half unit circle and negative frequencies onto the lower half. The complete linear frequency axis is wrapped around the unit circle with $\omega_s/2$ + multiples of 2π being mapped onto $z = -1$ (Fig. 7.1). The left half s plane (the region $s = \sigma + j\omega$ with $\sigma < 0$) maps according to $z = e^{\sigma T}e^{j\omega T} = re^{j\omega T}$ with $r < 1$ onto the interior of the unit circle in the z plane. Likewise, the right half s plane is mapped onto the outside of the unit circle.

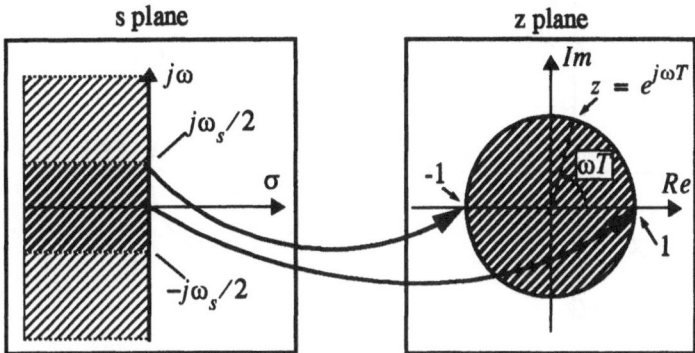

Fig. 7.1 : Mapping of the s plane to the z plane.

Just like the Laplace transform, equation (7.15) does not necessarily converge for all values of z. Those regions were it does exist are called regions of convergence. For continuous-time systems we have seen that we could directly relate most of the essential system properties to the distribution of poles and zeroes of the system in the s plane. The same is true for the discrete-time system. Moreover, we can directly transfer the relationships between pole-zero distribution in the s plane and system properties to the z plane if the singularities (poles and zeroes) are mapped according to $z = e^{sT}$ (Fig. 7.1). Singularities of transfer functions which lie in the left half s plane are mapped to singularities inside the unit circle in z while singularities in the right hand plane map to the outside of the unit circle.

Most of the transform properties for the z-transform could be directly guessed from their Laplace transform counterparts. About the two most important z-transform properties in the present context are the convolution theorem $(x_1[n] * x_2[n] \Leftrightarrow X_1(z) \cdot X_2(z))$ and the shifting theorem $(x[n - n_0] \Leftrightarrow z^{-n_0}X(z))$. In addition, it will become important in the next chapter that replacing z by $1/z$ in the z-transform $(X(z))$ corresponds to inverting the input signal $(x[n])$ in time $(x[-n] \Leftrightarrow X(1/z))$.

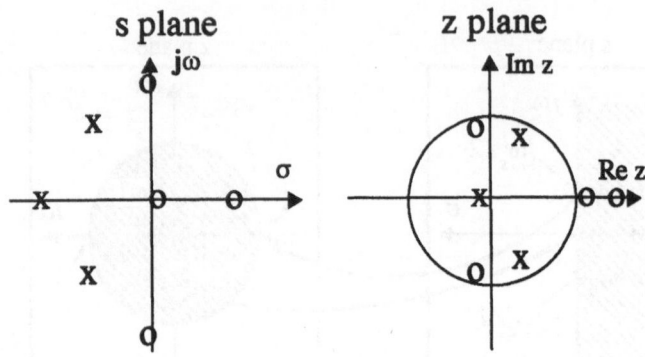

Fig. 7.2 Pole zero distribution in s plane and z plane. The poles in the s plane close to $j\omega$ will be mapped close to the unit circle. Poles on the real axis will stay on the real axis.

Hence from what we learned about the Laplace transform for continuous systems, we can intuitively justify the following properties for discrete systems:

• The region of convergence is a ring centered at the origin.

• The Fourier transform of $x[n]$ converges absolutely if and only if the unit circle is part of the region of convergence of the z-transform of $x[n]$.

• The region of convergence must not contain any poles.

• If $x[n]$ is a sequence of finite duration, the region of convergence is the entire z plane except possibly $z = 0$ or $z = \infty$ (see Problem 7.2).

• For a right-sided sequence the region of convergence extends outwards from the outermost pole to (and possibly including) $z = \infty$.

• For a left-sided sequence the region of convergence extends inwards from the innermost pole to (and possibly including) $z = 0$.

• For a two-sided sequence the region of convergence will consist of a ring bounded by poles on both the interior and the exterior.

These properties are further illustrated in Fig. 7.3.

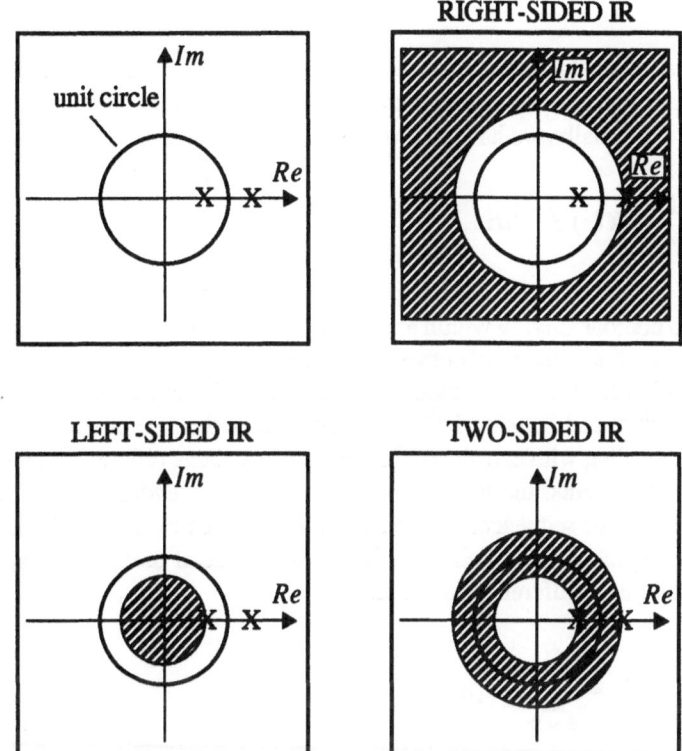

Fig. 7.3 Region of convergence and type of resulting impulse response functions (IR) for a discrete system containing two poles on the real axis (cf. Fig. 3.1).

Similar to the properties above, we can intuitively understand that for a discrete system to be causal and stable, all poles must lie inside the unit circle. If, in addition, the system contains no zeroes outside the unit circle, it is called minimum phase system.

Problem 7.1 Which type of impulse response function would result if we want to use the inverse Fourier transform for evaluation of the systems in Fig. 7.3?

Problem 7.2 How do time shifts effect the convergence of the z-transform for $z = 0$ and $z = \infty$? Argue by using the shifting theorem for z-transforms $(x[n - n_0] \Leftrightarrow z^{-n_0} X(z))$ for positive and negative n_0.

7.4 The inverse z-transform

When we need to evaluate the inverse z-transform, e.g. in order to obtain the impulse response function for a given z-transform transfer function we can proceed in several ways. Formally, we have to evaluate a contour integral

$$x[n] = \frac{1}{2\pi j} \oint_C X(z) z^{n-1} dz$$

(7.16)

with the closed contour C lying within a region of convergence and being evaluated counterclockwise. If the region of convergence includes the unit circle and if (7.16) is evaluated on it, the inverse z-transform reduces to the inverse Fourier transform (e.g. see Oppenheim and Schafer, 1989). However, since the type of discrete linear systems we are dealing with can always be described by linear difference equations with constant coefficients, the transfer functions will be a rational function in z. This is always true for sequences which can be represented by a sum of complex exponentials (Oppenheim and Schafer, 1989). For a general discrete LTI system described by the linear difference equation:

$$\sum_{k=0}^{N} a_k y[n-k] = \sum_{k=0}^{M} b_k x[n-k]$$

(7.17)

the discrete transfer function is defined as the z-transform of the output $Z\{y[n]\}$ divided by the z-transform of the input $Z\{x[n]\}$. Taking the z-transform of (7.17), we obtain using the shifting theorem $(x[n-n_0] \Leftrightarrow z^{-n_0} X(z))$

$$\sum_{k=0}^{N} a_k z^{-k} Y(z) = \sum_{k=0}^{M} b_k z^{-k} X(z)$$

(7.18)

and

$$H(z) = \frac{Y(z)}{X(z)} = \frac{\sum_{k=0}^{M} b_k z^{-k}}{\sum_{k=0}^{N} a_k z^{-k}} = \left(\frac{b_0}{a_0}\right) \frac{\prod_{k=1}^{M} (1 - c_k z^{-1})}{\prod_{k=1}^{N} (1 - d_k z^{-1})}$$

(7.19)

with c_k being the non-zero zeros and d_k being the non-zero poles of $H(z)$. Keeping in mind the correspondence between time shift by k samples in time and the multiplication with z^{-k} of the corresponding z-transform (shifting theorem), we can see that the coefficients of the polynomials in the transfer function $H(z)$ are identical to the coefficients of the difference equations. Hence, we can construct one from the other. Furthermore, for rational transfer functions we can use a partial fraction expansion to separate $H(z)$ into simpler terms for which the inverse z-transforms are tabulated and can be used directly.

7.5 The Discrete Fourier Transform (DFT)

During the discussion of the Fourier transform of discrete signals (sequences), we were assuming that the sequences we dealt with were of infinite duration. In reality of course, we are always dealing with finite duration sequences for which a different Fourier representation - the Discrete Fourier Transform (DFT) - is commonly used. In contrast to the Fourier transform for infinite sequences, it is not a continuous function but a finite length sequence itself. It corresponds to the Fourier series representation of the infinite periodic sequence which is made up by periodic extension of the given finite sequence $x[n]$. For a finite sequence $x[n]$ of length N, the DFT is defined as:

$$\tilde{X}[k] = \sum_{n=0}^{N-1} \tilde{x}[n] e^{-j2\pi kn/N} \tag{7.20}$$

Here $\tilde{x}[n]$ is the (infinite) periodic sequence constructed from $x[n]$ by periodic continuation. We can view the sequence $x[n]$ as being obtained by sampling a continuous time signal using a sampling interval of T. The DFT is only defined for discrete frequencies ω_k, which are related to the total number of points N, and the sampling interval T by:

$$\omega_k = k \cdot 2 \frac{\pi}{TN} \quad \text{for } k = 0, 1, \ldots - \ldots, N-1 \tag{7.21}$$

You can see from equation (7.20), that the DFT is a periodic function with a period of N. We should keep in mind that the Fourier transform for an infinitely long discrete sequence was also periodic, but still defined at continuous frequency values which is no longer the case for finite discrete sequences. If the sampling was done in agreement with the sampling theorem, the values of one period of the DFT are related to the values of the Fourier transform by:

$$X(j\omega)\big|_{\omega = \omega_k} = T \cdot \tilde{X}[k] \tag{7.22}$$

As it is sketched in Fig. 7.4, the DFT for $x[n]$ samples the z-transform of $\tilde{x}[n]$ at N equally spaced points on the unit circle.

z plane

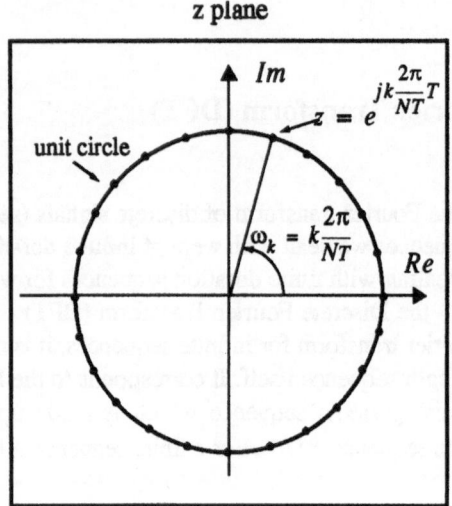

Fig. 7.4 : DFT representation in the z plane.

The set of sampled values can be recovered from the DFT by means of the inverse DFT, given by:

$$\tilde{x}[n] = \frac{1}{N}\sum_{k=0}^{N-1}\tilde{X}[k]\,e^{j2\pi kn/N} \tag{7.23}$$

If the length of the sequence equals an integer power of 2, the computation of the DFT is normally done using the time efficient FFT (Fast Fourier Transform) algorithm. If not, the sequences are normally padded with zeroes up to the next integer power of 2.

Similar to the continuous case, the discrete convolution of two sequences $x_1[n]$ and $x_2[n]$ is defined as

$$x_1[n] * x_2[n] = \sum_{m=-\infty}^{\infty} x_1[m]\, x_2[n-m] \tag{7.24}$$

Again, the output of a discrete filter can be calculated from the convolution of the impulse response of the filter with the input signal.

By analogy with continuous-time signals, we would expect that convolution could be done again in the frequency domain by simply multiplying the discrete DFT spectra. This is true, however, only if we take special precautions. Otherwise the results may be not what we want as we will see in the following example.

First, let us create the impulse response functions of two filters which perform just a simple time shift. The corresponding impulse response functions simply consist of time shifted impulses. For a sampling rate of 100 Hz and a time shift of 5.5 sec, the impulse response function would consist of a spike at position 550. For a 6 sec delay filter, the spike would have to be located at position 600. The convolution of the two filters would consist of a time shifting filter with a delay of 11.5 sec. In Fig. 7.5, you can see what happens with the naive approach simply multiplying the DFT spectra of the individual finite discrete signals in order to obtain the DFT spectrum of the output signal. For the calculation of the FFT, 1024 points were used.

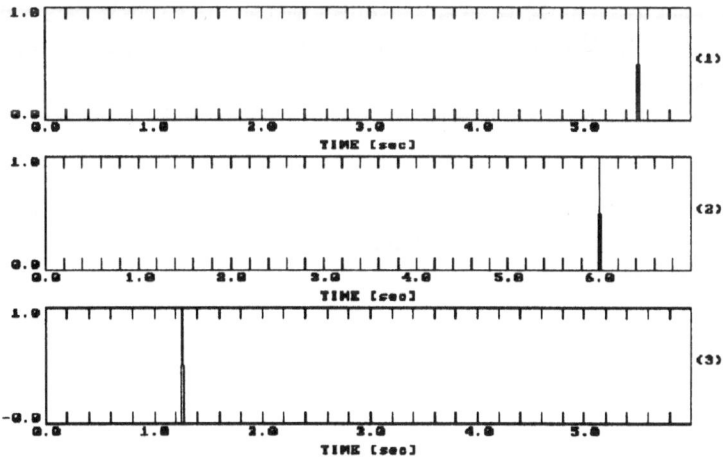

Fig. 7.5 The wrap around effect with discrete convolution in the frequency domain. Channel 3 is the inverse DFT of the product of the DFT spectra of channels 1 and 2. The no. of points used for calculating the DFT was equivalent to 10.24 sec.

From equation (7.24) we would expect that the output signal would have a spike at 11.5 sec (index 1150), which is more than we can display with 1024 points. Hence the output signal should be zero. Instead, we see that we have a spike at position 126. Well, we naively assumed that multiplying two DFT spectra gives the same result as multiplying two Fourier spectra. Instead, what we see is something called *wrap around effect* or *temporal aliasing*. This is due to the fact that for discrete systems the convolution theorem is slightly different

---

• *Convolution Theorem for discrete systems (circular convolution)* — If a sequence $\tilde{x}_1[n]$ is periodic with a period N, then its discrete convolution with another signal $\tilde{x}_2[m]$ of duration N is identical to the inverse DFT of the product of the corresponding DFT spectra.

Hence, naively multiplying two DFT spectra in order to obtain the same result as with discrete convolution will in general fail to produce the desired result. For N number of points in the spectra, the result will be the same as if the input sequence were periodic with the period of N. What we observe as wrap around effect is this periodicity. Likewise, the other properties of the DFT can be intuitively derived from the corresponding properties of the Fourier transform, only if we do not think in terms of the finite sequence $x[n]$ but always in terms of $\tilde{x}[n]$ which contains infinitely many replicas of $x[n]$.

When we want to avoid the wrap around effect we can use a simple trick which is called *zero padding*. If we artificially increase the period (N) by padding the signals with trailing zeroes to make it larger than the largest time lag for which the input signal could possibly be affected by the impulse response, no wrap around will occur.

Problem 7.3 How many points for zero padding would be needed for the example 4.1 in order to get rid of the wrap around effect?

8 The Digital Anti-Alias Filter

High performance seismic recording systems that make use of oversampling and decimation techniques relax the requirements on the analog anti-alias filters while performing most of the filtering in the digital domain. Since the resolution of such systems depends directly on the oversampling ratio, digital anti-alias filters with very steep transition bands are needed to gain the most from a given decimation ratio (cf. Fig. 6.9). On the other hand, the filter should leave a band-limited signal falling completely within the passband as unaffected as possible causing neither amplitude- nor phase distortions.

In order to implement digital filters, we can proceed in different ways. In terms of the concepts we have been encountered so far with general LTI systems, we have seen that a filter process can be described by the multiplication of a 'spectrum' with a 'transfer function'. Since right now we are dealing with discrete filters only, 'spectrum' in the most general sense would mean the z-transform of the discrete signal and 'transfer function' would refer to the z-transform of the discrete impulse response. Evaluating the z-transform on the unit circle, for filtering we would have to multiply the discrete Fourier spectrum of the signal (DFT) with the discrete frequency response function of the filter. Of course, we would have to worry about wrap around effects as discussed above. On the other hand, we could perform the filtering process in the time domain by directly evaluating the convolution sum. We could also make use of the fact, that the discrete systems we are dealing with can all be described by rational transfer functions in z (7.19)

$$H(z) = \frac{\displaystyle\sum_{k=0}^{M} b_k z^{-k}}{\displaystyle\sum_{k=0}^{N} a_k z^{-k}} = \left(\frac{b_0}{a_0}\right) \frac{\displaystyle\prod_{k=1}^{M} (1 - c_k z^{-1})}{\displaystyle\prod_{k=1}^{N} (1 - d_k z^{-1})} \tag{8.1}$$

with the corresponding linear difference equation (7.17):

$$\sum_{k=0}^{N} a_k y\,[n-k] \;=\; \sum_{k=0}^{M} b_k x\,[n-k] \tag{8.2}$$

relating the output sequence $y\,[n]$ to the input sequence $x\,[n]$. This concept naturally leads to two important classes of filters, *recursive* and *non-recursive filters*.
For recursive filters, the filter output $y\,[n]$, at a given sample n, not only depends
on the input values $x\,[n]$ at sample n and earlier samples n-k, but also on the output values $y\,[n-k]$ at earlier samples n-k, with k varying depending on N and M.
Hence with recursive filters, it is possible to create filters with infinitely long
impulse responses (IIR filters = Infinite Impulse Response filters). It should be
noted however, that not all recursive filters have to be IIR filters (cf. Hanning,
1977; example on p. 208). If $a_0 = 1$ and $a_k = 0$ for $k \geq 1$, $y\,[n]$ will depend
only on the input values $x\,[n]$ at sample n and earlier samples, n - k (for k running
from 0 to M) and the impulse response will be finite. These filters are called FIR
(Finite Impulse Response) filters. From equations (8.1) and (8.2), it can be seen
that the transfer function $H(z)$ of FIR filters is completely described by (c_k)
while that of IIR filters contains both poles and zeroes (c_k and d_k).

The filter design for given specifications could directly be done by placing poles
and zeroes in the complex z plane and exploiting what we have learned about system properties and the location of poles and zeroes of the corresponding transfer
function (cf. chapter "The z-transform" on page 80). For many situations, continuous filters with desirable properties may already be known or can be designed very
easily. For these cases, techniques are available to directly map continuous filters
into discrete ones. Some of these methods are based on sampling the impulse
response of a given filter under certain constraints (e.g. impulse invariance or step
invariance). Others, such as the bilinear z-transform, directly map the s-plane onto
the z-plane. Advantages and drawbacks of the individual approaches are discussed
in Oppenheim and Schafer (1989).

Important in the context of understanding modern seismic aquisition systems is the
fact that the kind of filter which is required for the digital antialiasing filter is most
easily implemented as a *FIR filter* as is discussed below. Although a complete discussion of the problem is beyond the scope of this text, some of the advantages and
disadvantages of FIR and IIR filters are listed below.

- <u>FIR filters</u>: They are always stable. For steep filters they generally need many coefficients, although part of this problem can be overcome by special design procedures. Filtering is fairly time consuming, especially for long filters. Both physically realizable (causal) and physically un-realizable (non-causal) filters can be implemented. Filters with given specifications are easy to implement.

- <u>IIR filters</u>: They are potentially unstable and subject to quantization errors (Since steep filters require poles and zeroes to be located close to but inside the unit circle, due to roundoff errors and the finite word length of the computer they may actually come to lie outside of the unit circle and causing potential unstabilities). They are always physically realizable (causal). Steep filters can easily be implemented with a few coefficients. As a consequence, filtering is fast. Filters with given specifications are in general hard, if not impossible to implement *exactly(!)*. [In order to create zero-phase responses from recursive filters, one can filter the signal twice in opposite directions which cancels the phase response. Obviously this can only be done with finite sequences, hence the name IIR filter would not be appropriate any more. Also, this approach can not be implemented as a recursive filter.]

The digital anti-alias filters most commonly employed in modern seismic recording systems belong to the class of generalized *linear-phase* filters which cause no phase distortions on the signal. In general, they cause a constant time shift which can easily be accounted for. If the time shift is zero or is corrected for, the filter would be called a *zero-phase* filter. To my knowledge, the reasons for using FIR filters in this context are primarily based on that well established design procedures exist to match given design specifications with great accuracy (McClellan et al., 1979), that very steep stable filters are needed, (these would be potentially unstable as IIR filters), that the linear phase property can be implemented exactly only as FIR filters, and most importantly, that IIR filters will always cause a phase distortion within the pass band of the filter.

In Fig. 8.1, the FIR filter impulse response is shown for the 8ms (125 Hz) data stream of the lennartz electronic MARS88 recording system.

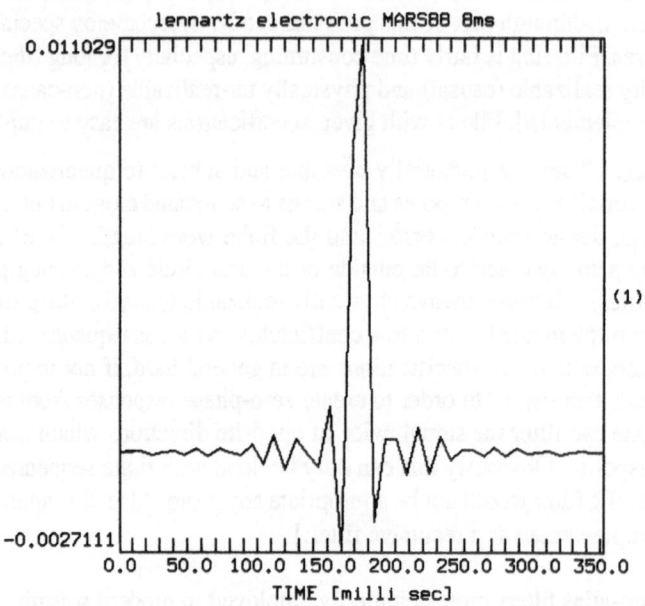

Fig. 8.1 FIR filter impulse response of the last stage of the lennartz electronic MARS88 recording system for an output sampling rate of 8ms (125 Hz). Before decimation to 125 Hz, the impulse response is perfectly symmetric.

These types filters with symmetrical impulse responses (before decimation) are often called two-sided or acausal for reasons discussed below. The oscillations of the impulse responses occur near the corner frequencies of the filters, an effect called Gibb's phenomenon. It could be decreased at the cost of the width of the filter pass band by decreasing the steepness of the filter. For seismological analysis, this effect may potentially obscure the onset of high frequency seismic signals. In Fig. 8.2, an example for this effect is shown for a record of a local earthquake in NE Bavaria, recorded at the station VIEL of the University of Munich seismic network. The trace shows the P wave portion on the vertical component decimated to 125 Hz using the FIR filter response shown in Fig. 8.1. As can be seen for this kind of high frequency signal, the use of the symmetric FIR filters may obscure the seismic onset of the phase considerably and may create a spurious acausality.

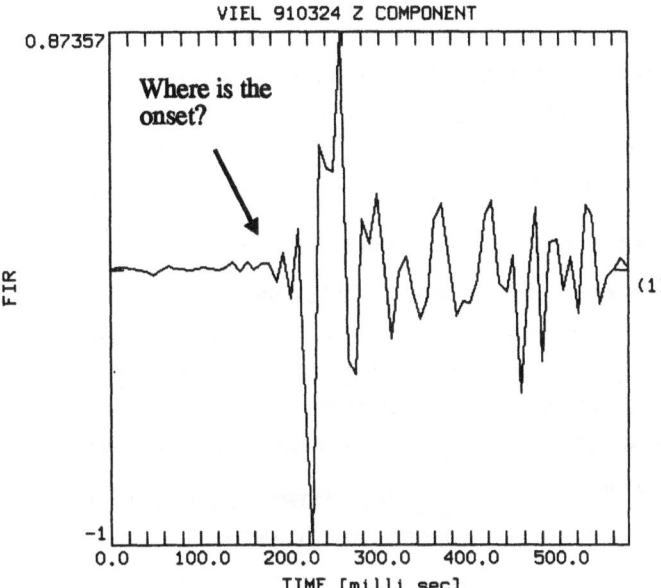

Fig. 8.2 Vertical component record of a local earthquake in NE Bavaria recorded at station VIEL of the University of Munich network using a MARS88 (lennartz electronic) recording system. The sampling rate is 8 ms (125 Hz). The digital anti-alias filters used is shown in Fig. 8.1.

One approach to deal with this problem which I will discuss below, is by constructing a filter which removes the acausal (left-sided) portion of the filtered signal and replaces it by its equivalent causal (right-sided) portion. Besides its own merits, this particular approach will provide a good opportunity to apply some of the concepts we have been dealing with so far.

In this context, a digital seismogram which is lowpass filtered by a general linear phase FIR filter (e.g., by the impulse response in Fig. 8.1) with additional correction for the constant time shift, can be viewed in terms of the z-transform as

$$\tilde{Y}(z) = F(z) \cdot z^{-lp} \cdot \tilde{X}(z) \tag{8.3}$$

Here, $\tilde{Y}(z)$ represents the z-transform of $\tilde{y}[n]$, the digitally lowpass filtered seismic trace before decimation, $F(z)$ the z-transform of the FIR filter, and z^{-lp} corresponds to a negative time shift by lp samples to make up for the linear phase component in $F(z)$. Hence $F(z) \cdot z^{-lp}$ corresponds to an acausal zero-phase filter.

8.1 Removing the acausal response of a FIR filter

If $f[n]$ is a general FIR filter, its z-transform $F(z)$ is stable in the complete z plane (Oppenheim and Schafer, 1989). Hence it only contains zeroes (cf. equ. (8.1)). These may lie inside and outside of the unit circle. The zeroes inside and on the unit circle (c_i^{min}) correspond to the minimum phase component $F_{min}(z)$, while the zeroes outside the unit circle (c_i^{max}) correspond to the maximum phase component of $F(z)$ which will be denoted $F_{max}(z)$. The zeroes outside the unit circle correspond to the left-sided ('acausal') portion of the impulse response (cf. chapter 7.3). In order to change the filter response of $F(z)$ to a causal one without changing the amplitude frequency response, we have to replace the maximum phase component $F_{max}(z)$ by its minimum phase equivalent, which will be denoted $MinPhase\{F_{max}(z)\}$. However, in order to change $F_{max}(z)$ into a minimum phase filter, we simply have to replace all zeroes lying outside the unit circle (c_i^{max}) by their reciprocal values ($1/c_i^{max}$). This, however, corresponds to simply inverting the impulse response of $F_{max}(z)$ in time, hence $MinPhase\{F_{max}(z)\} = F_{max}(1/z)$. This is one of the z-transform properties like the convolution theorem and/or the shifting theorem which really make one sometimes appreciate life in the z plane.

The process to make the response of $F(z)$ causal consists of deconvolving the maximum phase part and convolving with the equivalent minimum phase part. In terms of the z-transform this can be expressed by division and multiplication

$$Y(z) = \frac{1}{F_{max}(z)} \cdot F_{max}(1/z) \cdot \tilde{Y}(z) \qquad (8.4)$$

Now we have a problem! Since $F_{max}(z)$ has only zeroes outside the unit circle, $1/F_{max}(z)$ will contain only poles outside the unit circle for which only the left-sided impulse response is stable (cf. Fig. 7.3). However, if we invert all signals in time before filtering

$$Y(1/z) = \frac{1}{F_{max}(1/z)} \cdot F_{max}(z) \cdot \tilde{Y}(1/z) \qquad (8.5)$$

the impulse response corresponding to $1/F_{max}(1/z)$ becomes a stable (!) causal sequence in nominal time[1]. Hence, in nominal time, filtering with the anti-causal signal $F_{max}(z)$ poses no problems.

8.1.1 The difference equation

Rewriting equation (8.5) we obtain

$$Y(1/z) \cdot F_{max}(1/z) = F_{max}(z) \cdot \tilde{Y}(1/z) \tag{8.6}$$

which can be written as

$$A'(z) \cdot Y'(z) = B'(z) \cdot X'(z) \tag{8.7}$$

Here, $Y'(z)$ and $X'(z)$ correspond to $Y(1/z)$ and $\tilde{Y}(1/z)$, while $A'(z)$ and $B'(z)$ correspond to $F_{max}(1/z)$ and $F_{max}(z)$, respectively. Written as convolution sum, this becomes

$$\sum_{k=-\infty}^{\infty} a'[k] \cdot y'[i-k] = \sum_{k=-\infty}^{\infty} b'[k] \cdot x'[i-k] \tag{8.8}$$

If we assume $F(z)$ contains mx zeroes outside the unit circle, the wavelets $a'[n]$ and $b'[n]$ will be of length $mx+1$. Hence, equation (8.8) becomes

$$\sum_{k=0}^{mx} a'[k] \cdot y'[i-k] = \sum_{k=0}^{mx} b'[k] \cdot x'[i-k] \tag{8.9}$$

or

$$y'[i] \cdot a'[0] + \sum_{k=1}^{mx} a'[k] \cdot y'[i-k] = \sum_{k=0}^{mx} b'[k] \cdot x'[i-k] \tag{8.10}$$

which can be written as

---

1. This trick was suggested by E. Wielandt (pers. comm.,1993).

$$y'[i] = -\sum_{k=1}^{mx} \frac{a'[k]}{a'[0]} \cdot y'[i-k] + \sum_{k=0}^{mx} \frac{b'[k]}{a'[0]} \cdot x'[i-k] \qquad (8.11)$$

which is formally identical to equation (8.2). With

$$a[k] = -\frac{a'[k]}{a'[0]} = \frac{f_{max}[mx-k]}{f_{max}[mx]} \text{ for } k = 1 \text{ to } mx \qquad (8.12)$$

and

$$b[k] = \frac{b'[k]}{a'[0]} = \frac{f_{max}[k]}{f_{max}[mx]} \text{ for } k = 0 \text{ to } mx \qquad (8.13)$$

equation (8.8) becomes

$$y'[i] = \sum_{k=1}^{mx} a[k] \cdot y'[i-k] + \sum_{k=0}^{mx} b[k] \cdot x'[i-k] \qquad (8.14)$$

Hence, if we know the mx coefficients of $f_{max}[k]$, the maximum phase portion of a linear phase FIR filter, we can employ equation (8.14) to calculate $y'[i]$, the time reversed sequence for which the non-causal part of the FIR filter has been replaced by its equivalent minimum phase part. After filtering, we can obtain $y[i]$ by simply flipping $y'[i]$ back in time. Finally (cf. eq. (8.4)), $Y(z)$ can be written as

$$Y(z) = \tilde{F}(z) \cdot \tilde{Y}(z) \qquad (8.15)$$

$\tilde{F}(z)$ is a time advanced minimum phase filter advancing the output signal by lp samples, which of course can be easily corrected for.

8.1.2 Implementation in C

So far, we have never been concerned with the details of a specific algorithm. Here, we will make an exception, since this will allow us to discuss some problems commonly arising when trying to implement a specific signal processing technique. In the following I will be briefly discussing the program mkcausal.c which eliminates the noncausal part of the response of an arbitrarily asymmetric FIR filter by a direct implementation of equation (8.14). As a practical example for the reader, the source code listing is given in Appendix B. It can also be obtained through Internet directly from the author[1].

Prerequisite for all attempts to eliminate the noncausal part of a FIR filter response is the determination of its maximum phase component. In mkcausal.c, I have made use of a polynomial rooting technique to first determine the distribution of zeroes of $F(z)$ using the eigenvalue method described in Press et al. (1992). In comparison to techniques such as cepstrum, it has the advantage that it yields both $F_{max}(z)$ as well as mx, the length of the maximum phase portion of an arbitrarily asymmetric FIR filter. The disadvantage of this technique, however, is that it can only be used with relatively short FIR filter responses (cf. Scherbaum et al., 1994).

In order to get at the maximum phase portion a FIR filter of length $m+1$ we use a rooting technique to find the roots c_i^{max} ($i = 1$ to mx) of its z transform $F(z)$ which lie outside the unit circle. Next, we can rewrite $F_{max}(z)$ as product of its roots

$$F_{max}(z) = f[m] \prod_{i=1}^{mx} (z - c_i^{max}) \qquad (8.16)$$

with $f[m]$ being the last FIR filter coefficient. The corresponding anticausal impulse response can be obtained by evaluating (8.16) on the unit circle which is equivalent to calculating the inverse DFT for $F_{max}(e^{j\omega})$. The values for $f_{max}[n]$ can then directly be plugged into the difference equation (8.14) and used for filtering. In principle, this is how mkcausal.c works.

However, with the numerical implementation we are facing two common, sometimes confusing problems:

First, rooting algorithms such as zrhqr.c (Press et al., 1992) will provide the roots of polynomials $P(x)$ with $P(x) = \sum_{i=0}^{m} a_i x^i$. In contrast, however, with the definition

---

1. frank@bavaria.geophysik.uni-muenchen.de

of the z-transform which we have been using so far (7.15), $F(z) = \sum_{k=0}^{m} f[k] \cdot z^{-k}$
will only contain negative powers of k.

Secondly, the DFT implemented in Press et al. (1992) calculates
$\tilde{X}[k] = \sum_{n=0}^{N-1} \tilde{x}[n] e^{j2\pi kn/N}$ while we have been using a negative exponent in our
definition in equation (7.20).

We could deal with the first problem by simply multiplying $F(z)$ by z^m which
would correspond to a negative time shift of $f[n]$ by m samples and would introduce additional zeroes at the origin (see Problem 7.2). Because of the symmetry
between DFT and inverse DFT, we could deal with the second problem by using
the code for the DFT from Press et al. (1992) to calculate - after appropriate scal-
ing- the inverse DFT according to the definition we are used to. However, there is
an easier way out which solves both problems simultaneously, if we temporarily
use a different definition of the z transform

$$\chi(z) = \sum_{n=-\infty}^{\infty} x[n] \cdot z^n \tag{8.17}$$

With this definition, we can directly use $x[n]$ as input for the rooting algorithm.
Furthermore, the evaluation of the z-transform $\chi(z)$ on the unit circle directly corresponds to the evaluation of the inverse DFT according to Press et al. (1992).
What becomes different, however, are the relationships between the phase properties and the location of poles and zeroes. Using the definition of equation (8.17),
minimum phase filters must have all their poles and zeroes outside the unit circle
while maximum phase filters have their singularities located inside the unit circle.
For the FIR filter response of Fig. 8.1, the distribution of roots in the z plane using
the convention of equ. (8.17) is shown in Fig. 8.3.

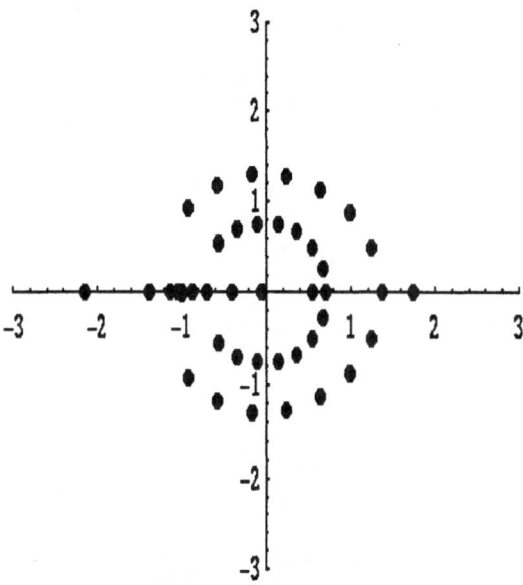

Fig. 8.3 Distribution of roots for the MARS88 FIR filter response shown in Fig. 8.1.

The performance of the program mkcausal on the FIR filter response of the MARS88 decimation filter shown in Fig. 8.1 is displayed in Fig. 8.1.

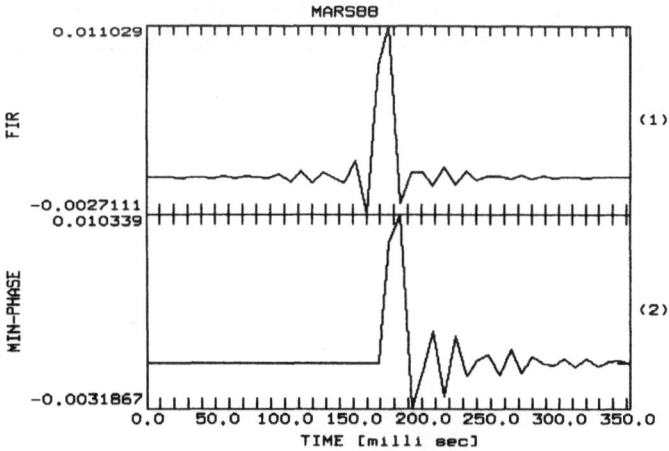

Fig. 8.4 Using mkcausal to remove the acausal part of the impulse response of the MARS88 decimation FIR filter shown in Fig. 8.1. Trace 1 shows the two-sided FIR filter response, trace 2 shows the filter response after making it causal.

In Fig. 8.5, the application of the correction filter to the data trace in Fig. 8.2 is shown. Notice the removal of the precursory signals before the P wave onset.

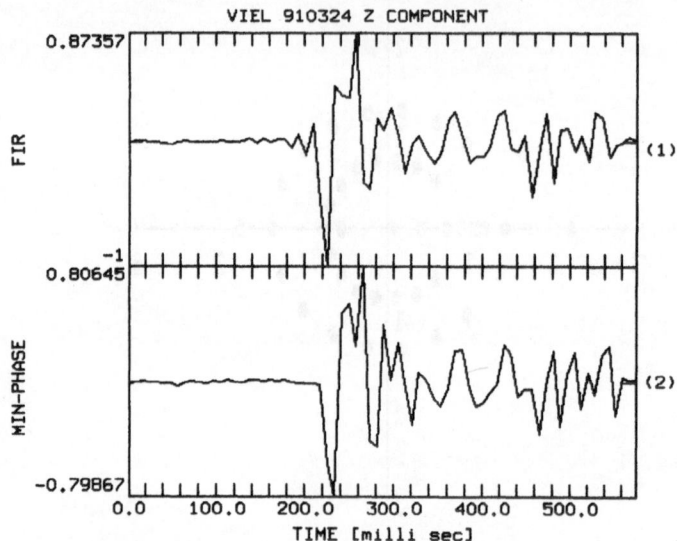

Fig. 8.5 Removing the left-sided (acausal) filter response from the signal of trace 1 of Fig. 8.2. The top trace shows the FIR filtered trace. The bottom trace shows the result of removing the acausal portions of the FIR filter response.

The correction filter to remove the acausalities is a pure allpass filter which leaves the amplitude portion of the frequency response function unaffected, however causing the phase to become minimum phase. For the correction filter used in Fig. 8.5, the frequency response in amplitude, phase and the group delay is displayed in Fig. 8.6.

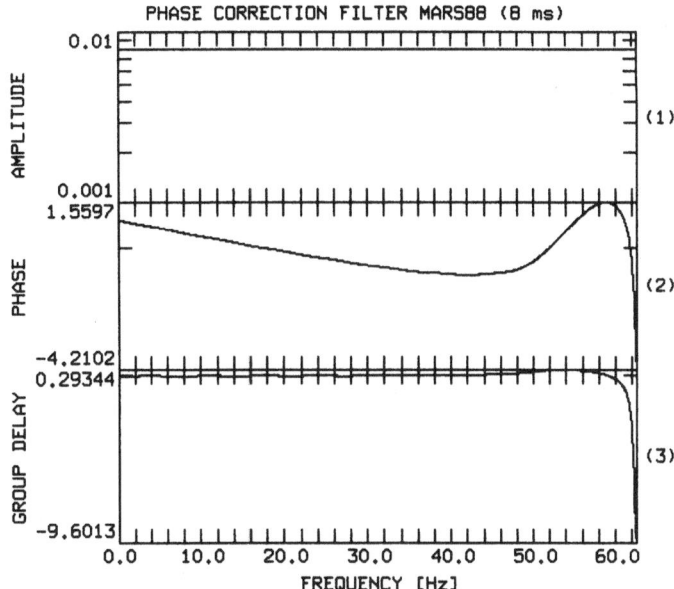

Fig. 8.6 Frequency response function of the correction filter used in Fig. 8.5. Displayed from top to bottom are the amplitude response, the phase response, and the group delay $(-d\Phi(f)/df)$. Note that the amplitude response is constant, hence the filter is a pure all-pass filter.

So far we have only considered FIR filters with zeroes away from the unit circle. While this turns out to be the case for the example given in Fig. 8.1, this is not necessarily in general the case. In contrary, symmetry conditions for certain FIR filter responses even require at least one root on the unit circle (Oppenheim and Schafer, 1989). The effects of roots on the unit circle on the properties of a FIR filter impulse response and their implication for removing the acausal filter response from seismic records is discussed in detail in Scherbaum et al. (1993). In the present context, I will leave it to the reader to consider the following problem.

Problem 8.1 Do roots on the unit circles belong to the minimum- or the maximum phase part or the impulse response? Start from reasoning by considering a short 3 point wavelet with roots exactly on the unit circle.

9 Deconvolution, a glimpse ahead

By now we have looked at all essential building blocks of modern seismic recording systems (cf. Fig. 6.13) and seen how they may affect the waveforms of
recorded seismic signals. We have learned how the individual components (seismometer, ADC, etc.) can be modelled within the concepts of continuous and discrete linear time invariant systems (transfer function, frequency response function,
impulse response function) and how there interaction can be described by convolution both in the time and frequency domain (convolution theorem). We have
become acquainted with the elegant and powerful concept of poles and zeroes and
learned what they can tell us about important system properties such as minimum
phase, maximum phase, etc. In addition, we have seen how we can use poles and
zeroes for the design of some simple special purpose digital filters. After having
brushed up on the theoretical background of Fourier-, Laplace-, and z-transforms,
we have taken a detailed look at the digital anti alias filter. We have seen how we
can come up with a cure for the precursory ringing problem of high frequency signals and discussed the design and implementation of a recursive correction filter.

So far we have been mainly concerned with the *forward problem*, the modelling of
an output signal for arbitrary input signals and known system properties. Although
the in depth treatment of the corresponding *inverse problem* - the determination of
the input signal from a given output signal and known signal properties (*deconvolution*) - is beyond the scope of this course on "basic concepts", it is of major
importance in the context of digital signal processing in seismology. For some of
the practical aspects of this topic see Seidl (1980), Seidl and Stammler (1984) and
Plesinger and Zmeskal (1993).

With the final examples in this course, I want to take an naive approach and see
what problems are going to be encountered in that context. From the convolution
theorem we know how to describe the interaction of a general linear time invariant
filter with an arbitrary input signal by multiplication of the appropriate spectra.
Hence, for the deconvolution of the impulse response of a recording system, we
can - with the proper zero padding - calculate the DFT spectra of the seismic trace
and divide it by the DFT spectrum of the impulse response and ... would possibly
be surprised. The reason is that the spectral division enhances those signal compo-

nents for which the frequency response function of the instrument has very small values. Fig. 9.1 shows the GRF A1 (EW component) recording of an M_{WA} = 4.9 earthquake below Frankfurt at a distance of about 180 km.

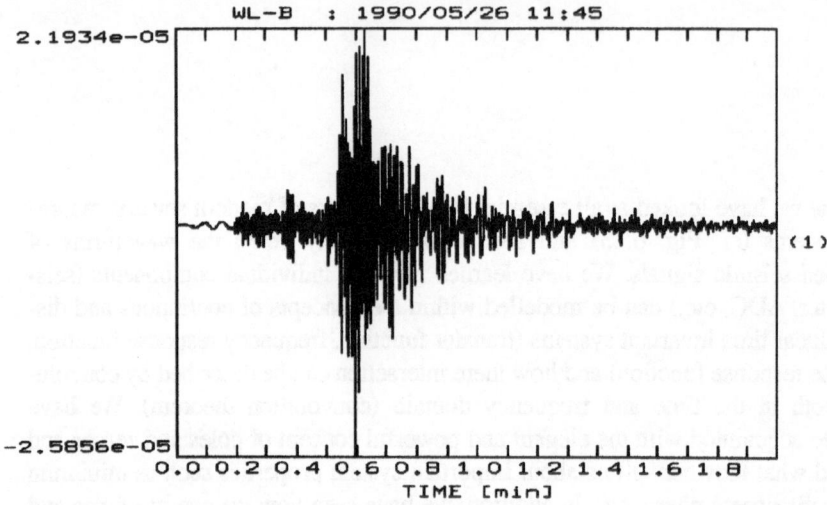

Fig. 9.1 GRF A1 (EW component) recording of an M_{WA} = 4.9 earthquake (1990 May 26) below Frankfurt.

The frequency response function of the recording system to ground velocity is shown in Fig. 9.2.

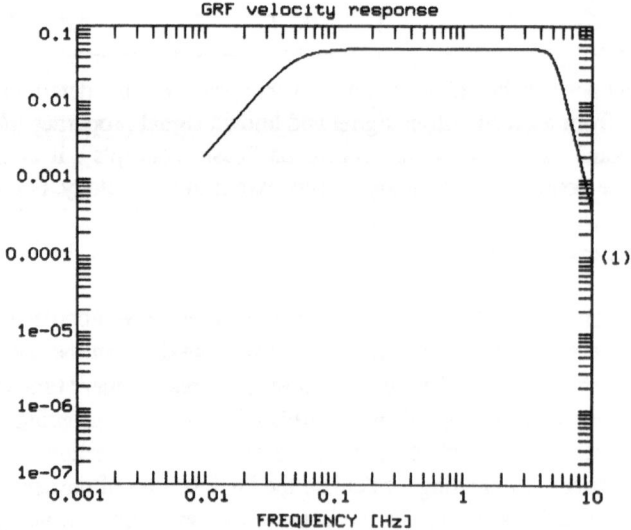

Fig. 9.2 GRF velocity frequency response function (amplitude).

The digitization frequency is 20 Hz. Notice the rolloffs of the frequency response function at 0.05 Hz (seismometer) and 5 Hz (anti-alias filter), respectively. Deconvolution by spectral division results in the trace shown inFig. 9.3.

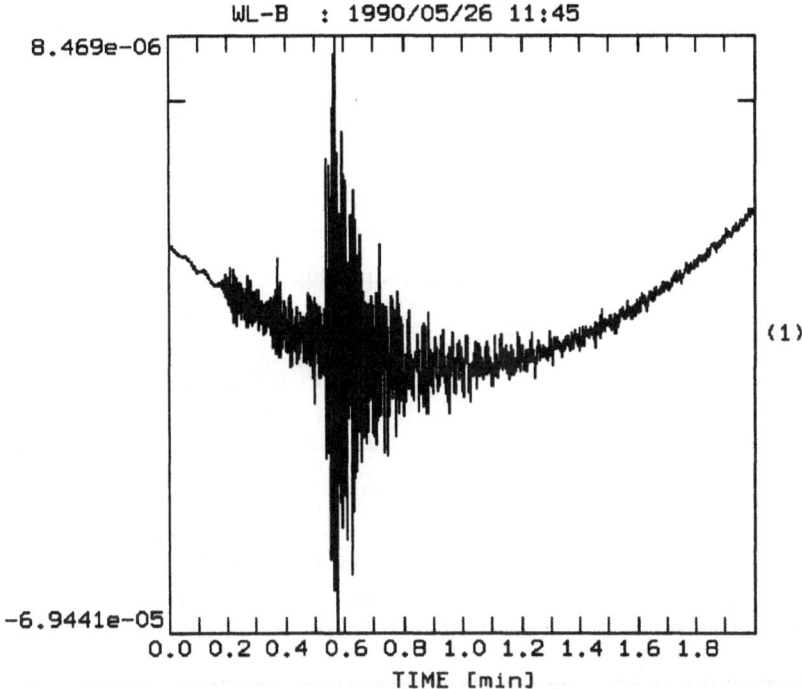

Fig. 9.3 Deconvolution by spectral division of the signal shown in Fig. 9.1using the frequency response function displayed in Fig. 9.2.

Although the deconvolution has not gone completely crazy, we can easily see that we are beginning to run into problems with stability. In Fig. 9.4 the deconvolution is visualized in the spectral domain. From top to bottom the amplitude spectrum of the observed trace, the amplitude part of the inverse frequency response function, and the amplitude spectrum of the deconvolved trace are shown, respectively. The bottom trace has been obtained as the complex product of the top and central trace. Fig. 9.4 shows how the low frequency and high frequency portions of the signal are disproportionally enhanced during the deconvolution process. As a consequence, the deconvolved signal may be dominated by these signal components, especially if they are corrupted by noise. One way to decrease the influence of this problem is to postfilter the deconvolved trace with a band pass filter selecting only the stable signal component.

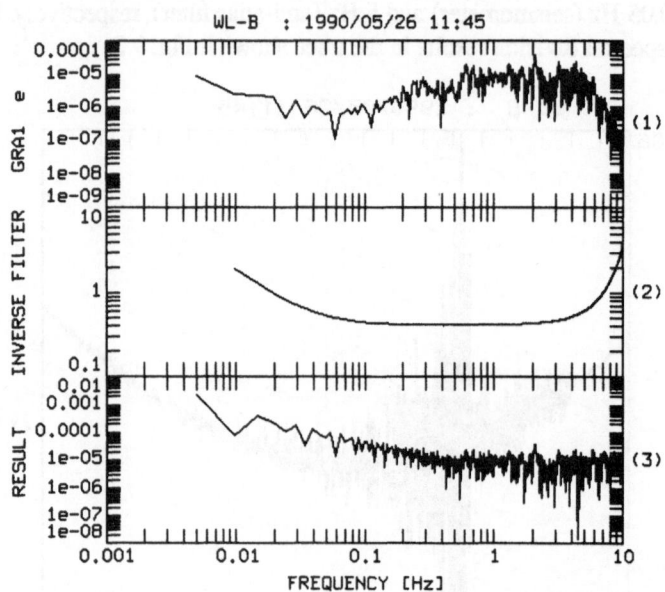

Fig. 9.4 Deconvolution in the frequency domain. From top to bottom, the amplitude spectrum of the observed trace, the frequency response function of the inverse filter, and the amplitude spectrum of the deconvolved trace are displayed.

Another method that is in common use is the so-called *water level* correction. It consists of enforcing a threshold for the amplitude values of the denominator spectrum while not changing the phase. The water level is measured from the maximum value of the amplitude spectrum. In other words, the dynamic range of the denominator spectrum is traded for the stabilization of the spectral division. There is no general rule for the selection of the water level. The denominator spectrum should be affected just enough to ensure stability of the deconvolution without sacrificing too much of the resolution. In Fig. 9.5, a waterlevel correction of 20 dB has been applied to the deconvolution of the signal in Fig. 9.1. Notice that this effectively removes the long period instability seen in Fig. 9.3.

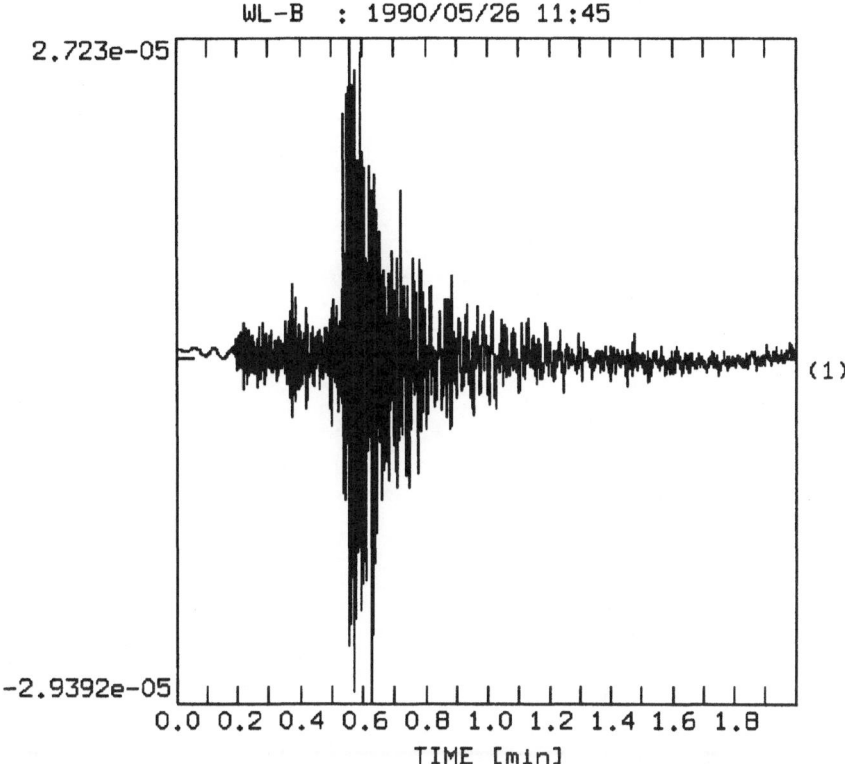

Fig. 9.5 Deconvolution by spectral division using a waterlevel stabilization of 20 dB of the signal shown in Fig. 9.1. The frequency response function is displayed in Fig. 9.2.

Besides the problem of stability, there are a number of other considerations to be taken into account. One of the most crucial ones for the measurement of onset times is the problem of spurious acausality which can arise in the context of deconvolution as well as instrument simulation. These topics, however, go beyond the aim of the present course and are treated in more specialized texts (e. g Plesinger and Zmeskal, 1993).

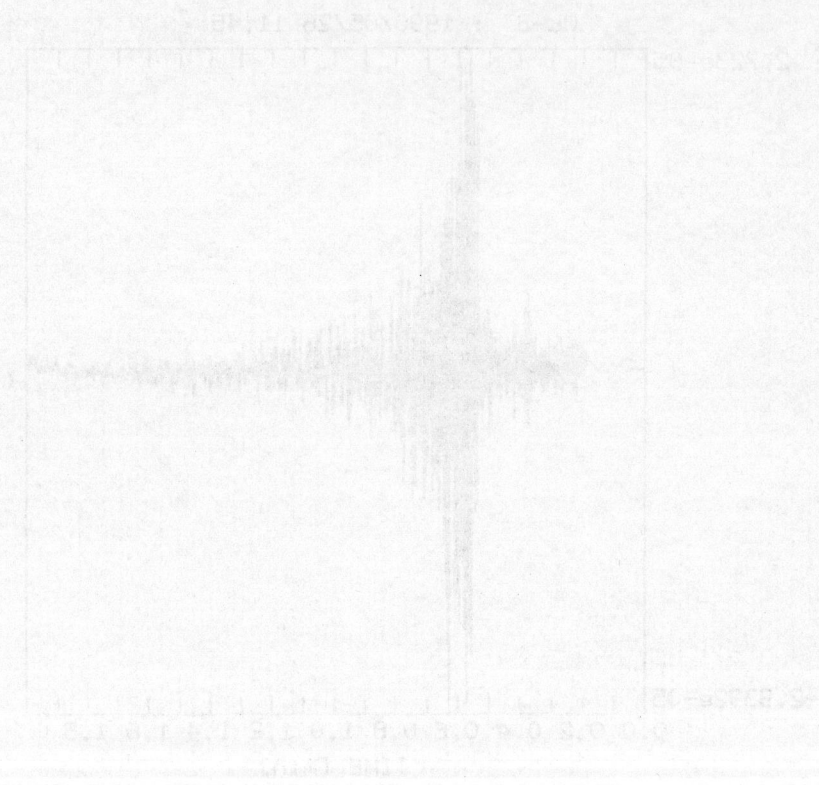

References

Dornboos, D. J., E. R. Engdahl, and T. H. Jordan, International Seismological Observing Period, Preliminary Science Plan, USGS, Box 25046, MS 967, Denver Federal Center, Denver, CO 80225, 1990.

Hamming, R. W., Digital filters, Sec. Edition, Prentice-Hall Signal Processing Series, A.V. Oppenheim (series editor), 257pp, 1977.

Hjelt, S. E, Pragmatic inversion of geophysical data, Lecture Notes in Earth Sciences 39, Springer-Verlag, 1992.

Jaeger, R.C., 1982., Tutorial: Analog data aquisition technology, part II - Analog to Digital Conversion, IEEE Micro, Vol. 2, August 1982, 46 - 56.

Luzitano, R. D., Data report for dense array recordings from nine aftershocks of the July 21, 1986 Earthquake in Chalfant Valley, California, USGS open file report 88-71, 1988.

McClellan, J. H., T. W. Parks, and L. R. Rabiner, FIR Linear phase filter design program. Programs for digital Signal Processing, IEEE Press New York, 1979.

Menke, W., Geophysical Data Analysis: Discrete Inverse Theory, Academic Press, 1984.

Oppenheim, A. V., and R. W. Schafer, Discrete-Time Signal Processing, Prentice Hall, 1989.

Plesinger, A., and M. Zmeskal, Manual on digital seismogram analysis, ISOP project, in prep. 1993.

Press, W. H., S. A. Teukolsky, W.T. Vetterling, and B. P. Flannery, Numerical Recipes in C, Second Edition, Cambridge University Press, 1992.

Scherbaum, F., Short Course on First Principles of Digital Signal Processing for Seismologists, IASPEI Software Library, Vol. 5, 1993.

Scherbaum, F., and J. Johnson, PITSA, Programmable Interactive Toolbox for Seismological Analysis, IASPEI Software Library, Vol. 5, 1993.

Scherbaum, F., E. Wielandt, and J. M. Steim, Removal of the noncausal FIR filter response from digital seismic records, in prep., 1994.

110

Seidl, D. The simulation problem for broad-band seismograms, J. Geophys., 48, 84-93, 1980.

Seidl, D., and W. Stammler, Restoration of broad-band seismograms (Part I), J. Geophys., 114-122, 1984.

Stearns, S. D., Digital Signal Analysis, Hayden Book Company, 1975.

Strum, R. D., and D. E. Kirk, First principles of discrete systems and digital signal processing, Addison-Wesley, 1988.

Tarantola, A., Inverse Problem Theory, Elsevier, 1987.

Appendix A: Solution to Problems

Chapter 2

Solution 2.1 The pole position is at $-1/\tau$. $\tau = R \cdot C = 4.0\Omega \cdot 0.1989495 F$ which is $4.0V/A \cdot 0.1989495 A\sec /V = 0.795798\sec$. Hence, the pole is at -1.2566 (rad/s). For each point on the imaginary axis (angular frequency axis), determine the reciprical of the length of the vector from the pole to that point. You can do this either by using a ruler and graph paper or simply by exploiting analytical geometry. Plot this value as a function of angular frequency or frequency, respectively.Below, the procedure is demonstrated schematically for a frequency of 1Hz (Fig. A 2.1; note, that Fig. A 2.1 is not on 1:1 scale).

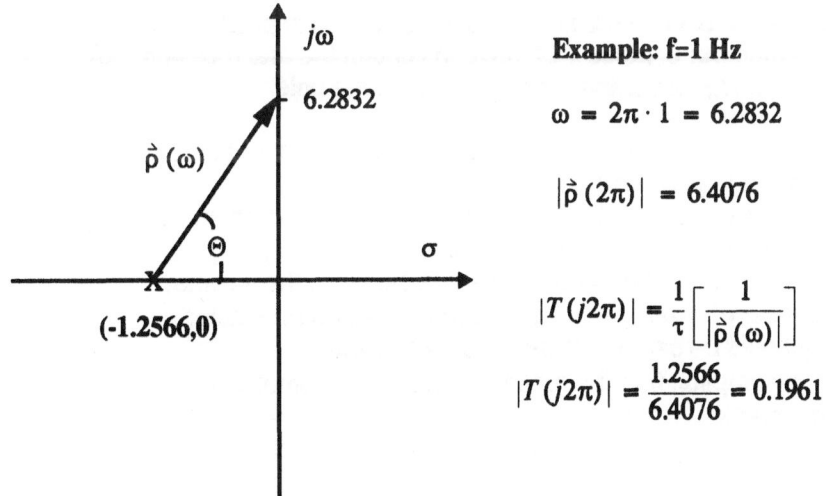

Example: f=1 Hz

$\omega = 2\pi \cdot 1 = 6.2832$

$|\vec{\rho}(2\pi)| = 6.4076$

$|T(j2\pi)| = \frac{1}{\tau}\left[\frac{1}{|\vec{\rho}(\omega)|}\right]$

$|T(j2\pi)| = \frac{1.2566}{6.4076} = 0.1961$

Fig. A 2.1 Graphical determination of the frequency response function (amplitude only) for the RC filter of Problem 2.1. The plot demonstrates the evaluation for a frequency of 1 Hz.

Solution 2.2 PITSA simulates the action of systems defined by their transfer function in the complex s plane. As we will see later in more detail, a transfer function of a more general system can have more than one pole as well as a number of zeroes (at which the transfer function becomes zero). The positions of poles and zeroes define the transfer function completely.

In order to do the filtering, PITSA needs to know the position(s) of the pole(s) and zeroe(s) in the complex s plane. They have to be provided in a simple ASCII file which can be created using your favourite text editor. In addition, you have to provide an input signal which we will now create using the 'test signal' tool in PITSA. From the list of poles and zeroes PITSA calculates the corresponding frequency response function. It then performs a complex multiplication of the frequency response function with the discrete Fourier spectrum of the input signal. Finally, it calculates the inverse Fourier transform. The whole process is equivalent to convolving the input signal with the impulse response of the system.

First you have to create an ASCII file containing the pole position in the complex s plane. This file has to conform to a special format (format of the calibration section for GSE data, see also the PITSA manual) which is also used in a different context:

line [1]: has to contain the string CAL1 in positions 1-4;
         character position 32 - 34 have to read PAZ.
line [2]: has to contain the number of poles (given in position 1-8)
line [3] - [number of poles + 2]: have to contain the real (position 1-8)
         and imaginary parts (position 9-16) of the pole
         positions in the s plane.
line [number of poles + 3]: has to contain the number of zeroes
         (given in position 1-8)
line [number of poles + 4] - [number of poles + 3 + number of zeroes]:
         have to contain the real (position 1-8) and imaginary
         (position 9-16) parts of the zero positions in the s plane.
line [number of poles + number of zeroes + 4]: scaling factor
         (should be set to 1.0E09 to correct for the
          assumption that the motion is given in nanometers).
last line: blank

For the RC-filter, the pole-zero file looks like:
CAL1   PAZ
1
-1.2566 0.0
0
1.0E09

Next, create one trace of a noise-free spike signal (to be found under utilities -> test signals) with a digitization frequency of 10.0 Hz and a trace length of 1024 points. For the time of the first sample, just accept the default values. Create a single spike at index 0 with an amplitude of 1.0 (enter a -1 for the second spike position). Accept/append the trace to the data in memory and perform a subsequent filtering using the pole-zero file option. Enter the name of the ASCII file containing the pole position (rcfilter.cal) and take the default values for digitization frequency, time of first sample and no. of points for the FFT. The signal you are seeing as a result of the filtering is the impulse response of the filter (since the input signal was an impulse). In order to see the frequency response function calculate a spectrum (FFT, untapered). Accept the zoom window as is and leave the default value for the number of points for FFT (1024). Choose amplitude for display. The plot should give the same result as Fig. 2.6.

Note: In case you put the spike not a zero position but at let's say position 512, you will notice some oscillations before the 'onset' of the filtered signal. This is due to the fact that the true spectrum of the filtered trace would not be zero at and above the Nyquist frequency (See "The sampling process" on page 57.). Since the spectrum for the filtered trace is calculated only up to 5 Hz, this corresponds to multiplying the true spectrum by a boxcar window in the frequency domain. This, however, is equivalent to convolving the causal impulse response with the *two-sided function* $\sin(at)/(2\pi at)$, the inverse Fourier transform of the frequency boxcar window ($F(j\omega) = 1/(2a)$ for $\omega \leq a$ and $F(j\omega) = 0$ for $\omega > a$, Fig. A 2.2).

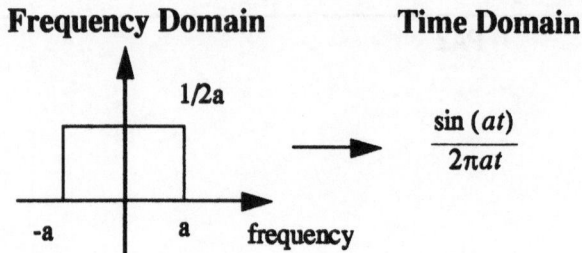

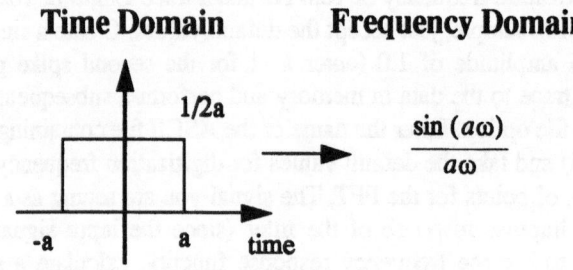

Fig. A 2.2 Boxcar windows in frequency and time domain.

Solution 2.3 The balance at time $t = nT + T$:

$$y(nT + T) = y(nT) + \alpha y(nT) + \alpha x(nT) + x(nT)$$

$$= y(nT) + \alpha y(nT) + (\alpha + 1) \cdot x(nT)$$

$$y(nT + T) - y(nT) = \alpha y(nT) + (\alpha + 1) \cdot x(nT) \qquad \text{(divide by T)}$$

$$\frac{y(nT + T) - y(nT)}{T} = \frac{\alpha}{T} y(nT) + \frac{(\alpha + 1)}{T} \cdot x(nT)$$

Next, replace the difference quotient by the derivative and nT by t:

$$\dot{y}(t) - \frac{\alpha}{T} y(t) = \frac{(\alpha + 1)}{T} \cdot x(t)$$

Taking the Laplace transform yields

$$s \cdot Y(s) - \frac{\alpha}{T} \cdot Y(s) = \frac{\alpha + 1}{T} \cdot X(s)$$

$$(s - \frac{\alpha}{T}) \cdot Y(s) = \frac{\alpha + 1}{T} \cdot X(s)$$

$$T(s) = \frac{Y(s)}{X(s)} = \frac{\alpha + 1}{T} \cdot \frac{1}{s - \frac{\alpha}{T}}$$

From the equation above we see that a pole exists at $s = \alpha/T$

The corresponding impulse response becomes: $\dfrac{\alpha + 1}{T} \cdot e^{\frac{\alpha}{T} \cdot t}$

From the positive exponent we see that the impulse response is unstable! While this is normally an unwanted feature, for a checking account it is just was we expect!

Chapter 3

Solution 3.1 The GSE format calibration files for the solution of Problem 3.1 are shown in Tables A3.1-A3.3.

Table A 3.1 GSE format pole-zero file for the solution of Problem 3.1a.

CAL1	PAZ
2	
-1.2566 0.0	
-1.2566 0.0	
0	
1.0E09	

Table A 3.2 GSE format pole-zero file for the solution of Problem 3.1b.

CAL1	PAZ
2	
-1.2566 0.0	
1.2566 0.0	
0	
1.0E09	

Table A 3.3 GSE format pole-zero file for the solution of Problem 3.1c .

CAL1	PAZ
2	
1.2566 0.0	
1.2566 0.0	
0	
1.0E09	

From the argument in chapter 3.1 'Generalization of concepts', the resulting impulse response function will be a) right sided, b) two-sided, and c) left sided as shown in Fig. A 3.1. The amplitude portion of the frequency response function is identical in all three cases while the phase response differs. With respect to Problem 2.2, the slope of the amplitude response is steeper by 20dB/decade since an additional pole is present.

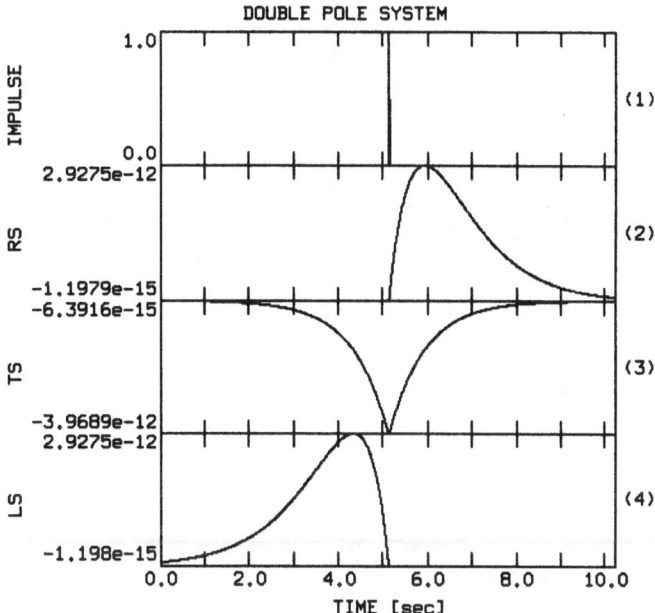

Fig. A 3.1 Impulse response functions for Problem 3.1.

118

Solution 3.2 Since the pole and the zero become symmetrical to the imaginary axis, the pole vectors and zero vectors are always of equal length. Hence their ratio is always constant. It follows that the amplitude response is the same for all frequencies, characteristic of an allpass filter (Fig. A 3.2).

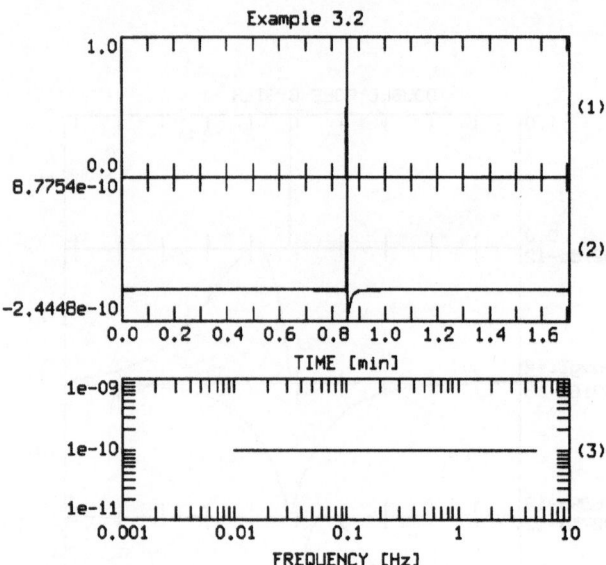

Fig. A 3.2 Input impulse, impulse response, and amplitude part of the frequency response function of the allpass filter of Problem 3.2.

Solution 3.3 a) A general transfer function F(s) will be described by poles and zeroes. Taking the inverse, 1/F(s) causes all poles to become zeroes and zeroes to become poles. For a minimum phase system, all the poles and zeroes of the transfer function F(s) are on the left half s plane. This will not change for the inverse system. Hence, the inverse system will also be minimum phase and therefore stable.
b)The poles and zeroes of a general mixed phase system with zeroes on either half plane can always be expanded into two systems. The allpass is constructed by taking all right half plane zeroes and adding symmetric poles. The minimum phase system is constructed by taking all poles (which have to be in the left half plane anyway because of stability) and adding zeroes where the allpass system had added poles for symmetry reasons. On multiplication of the two transfer functions these additional poles and zeroes will cancel (Fig. A 3.3).

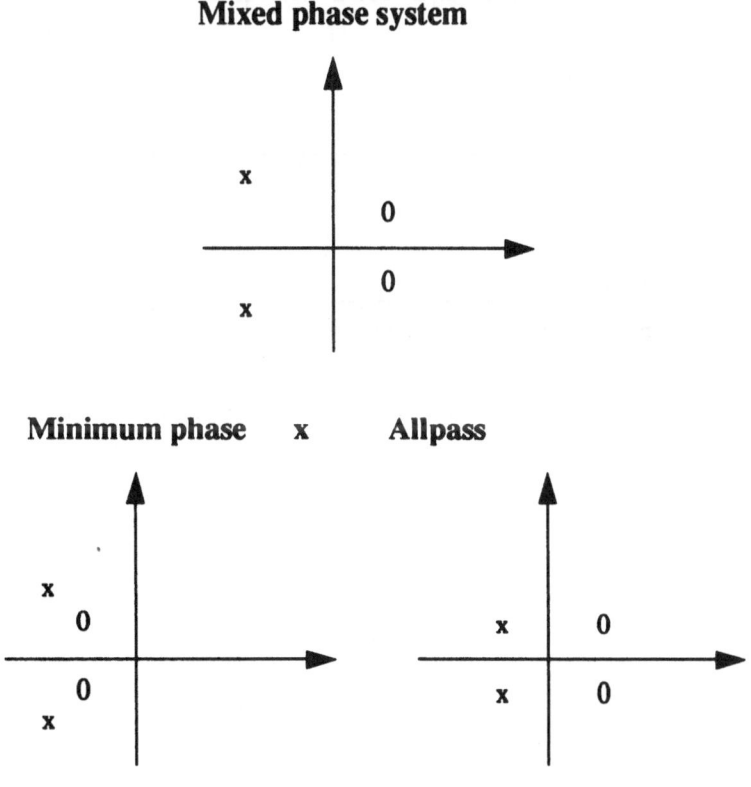

Fig. A 3.3 Separating a mixed phase system into a minimum phase system and an allpass.

120

Solution 3.4 Since the transfer function for problem 3.1b has a pole on either side of the imaginary axis, we need to cancel the right-sided pole by a zero. In order not to change the amplitude characteristic, we have to do this with an allpass filter whose poles and zeroes are symmetrical to the imaginary axis. The filter we need has to have a zero at +1.2566 and a pole at -1.2566, which is exactly the allpass from Problem 3.2. The performance on the trace from Problem 3.1b is shown in Fig. A 3.4.

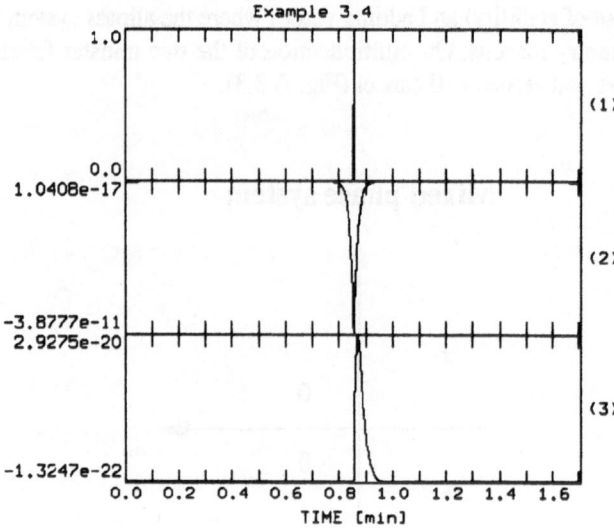

Fig. A 3.4 Changing the two sided impulse response of Problem 3.1b into a right sided one by application of an allpass filter.

Solution 3.5 Using the pole-zero file filter option in PITSA, we have to create the file shown in Table A 3.4. The amplitude part of the corresponding frequency response function is shown in Fig. A 3.5.

Table A 3.4 GSE format pole-zero file for the solution of Problem 3.5

CAL1	PAZ
1	
-6.28318 0.0	
1	
.628318 0.0	
1.0E09	

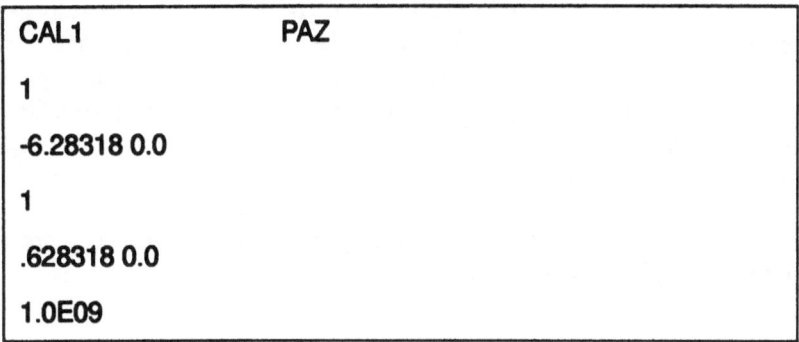

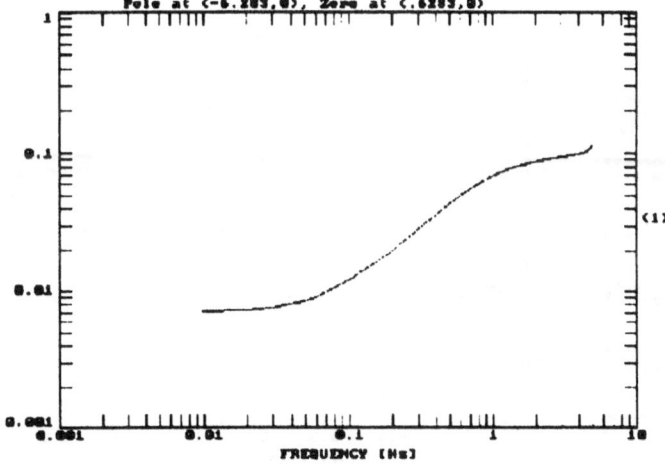

Fig. A 3.5 Amplitude part of the frequency response function for a system consisting a pole at (-6.28318, 0.0) and a zero at (.628318, 0.0).

From the pole and zero positions we expect to corner frequencies at $f_1 = 0.1$ Hz and at $f_2 = 1$ Hz, respectively. We can see that the frequency response function starts out constant, increases by 20 db/decade at f_1 then decreases by 20 dB/decade at f_2.

Solution 3.6 If you use PITSA, you could start out using the GSE format pole zero file in Table A 3.1 which represents the double pole at (-1.2566, 0). For the successive steps, the pole positions would be $p_{1,2} = -1.2566\,(\cos\alpha + i\sin\alpha)$

with α being incremented from 15° to 75° in steps of 15° (plus $\alpha = 85^o$, α is measured clockwise from the negative real s axis). The resulting impulse response functions are shown in Fig. A 3.6. In Fig. A 3.7, the corresponding amplitude portions of the frequency response functions are displayed. The closer the poles get to the imaginary axis, the more oscillations take place in the impulse response. Or if we look at it the other way around, the farther the poles move away from the imaginary axis, the stronger is the damping of the oscillations.

Notice that the spectral changes significantly only near the corner frequency, not at the high and low ends of the spectrum.

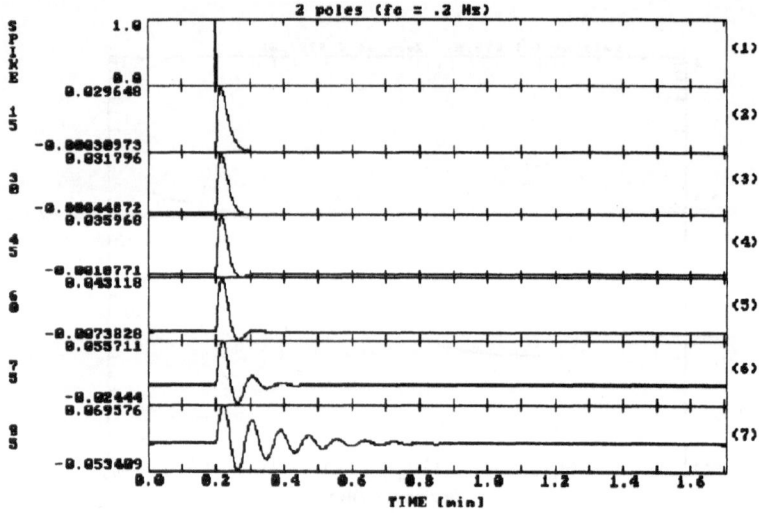

Fig. A 3.6 Impulse responses for a system with a conjugate complex pole with the pole positions at different angles from the real axis of the s plane (15, 30, 45, 60, 75, 85°).

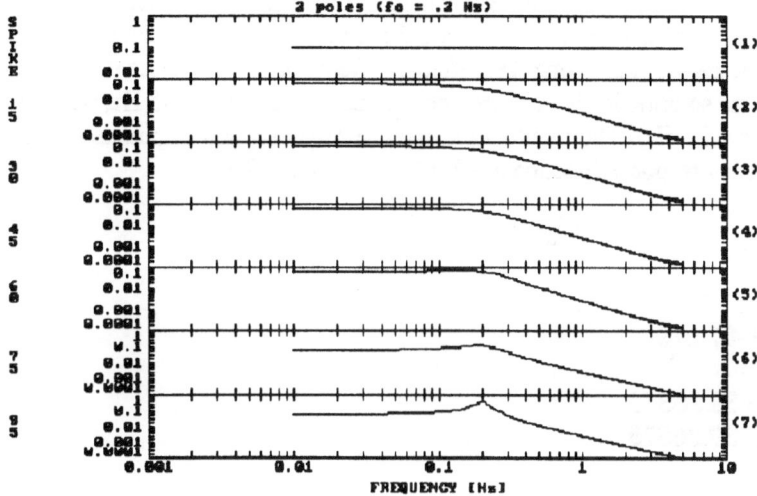

Fig. A 3.7 Frequency response functions (amplitude) corresponding to Fig. A 3.6.

124

Solution 3.7 In Problem 3.6 we have seen that the amplitude part of the frequency response functions shows a strong peak if the poles were close to the imaginary axis. Hence, zeroes close to the imaginary axis will correspond to strong selective suppression which is what we need. Therefore we need to put zeroes on or very close to the imaginary axis. In order to sharpen the notch, we also have to put poles close to the zeroes, cancelling their effect for frequencies away from the notch frequency. One solution is given below (pole corner frequency: 6.275 Hz, zero corner frequency: 6.25 Hz). The input spike, impulse response, and amplitude portion of the frequency response function are displayed in Fig. A 3.8.

CAL1   PAZ
2
-27.879 27.879
 -27.879 -27.879
2
 0.68525 39.26375
0.68525 -39.26375
1.0

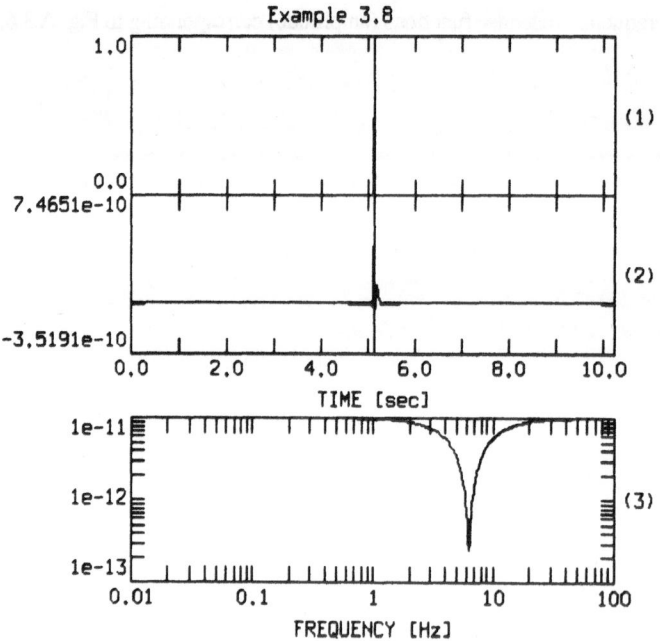

Fig. A 3.8 Notch filter at 6.25 Hz.

Solution 3.8 First estimate the different slopes and determine the number of different poles and zeroes which are needed to model them. One reasonable interpretation is sketched in Fig. A 3.9. Next, try to find the corner frequencies (0.5 Hz, and 20 Hz). For the transitional portion of the spectrum, you have to experiment with putting the poles at different distances from the imaginary axis.

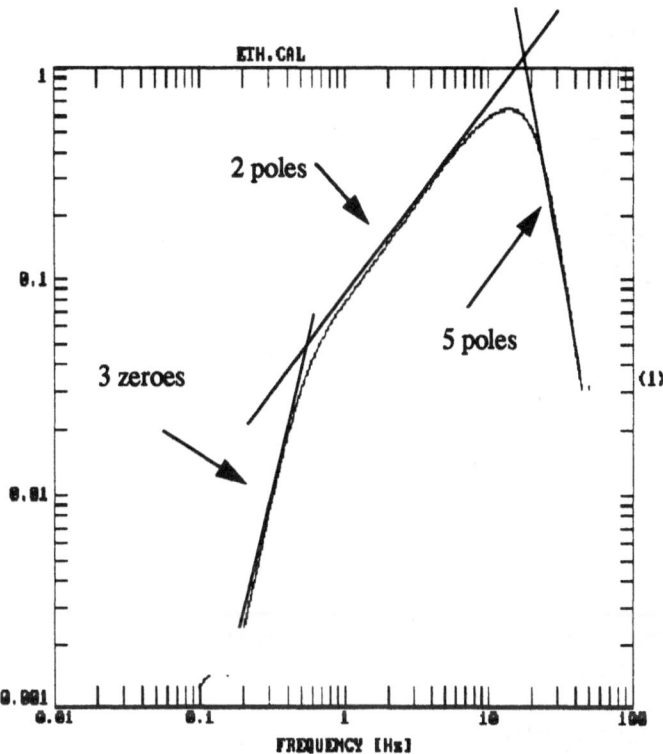

Fig. A 3.9 Frequency response function (amplitude) with an unknown pole - zero distribution.

The file from which this was constructed was:

```
XW01 CHE STO 000032 0339 900620 2117 C 000030 900619
CAL1 OSS   A700SZNG SZ MKIIIA PAZ 900101 0000
11
 -1.8850 2.5133
 -1.8850 -2.5133
-0.05610    0.0
-0.49474    0.0
-0.49474    0.0
-118.490 -24.198
-118.490 24.198
-104.181 -73.251
-104.181 73.251
-70.19 -125.30
-70.19 125.30
6
0.0    0.0
0.0    0.0
0.0    0.0
0.0    0.0
0.0    0.0
0.0    0.0
1.0
```

The corresponding corner frequencies are from top to bottom:
0.5 Hz (double pole)
0.0089 Hz (single pole)
0.07874 Hz (double pole)
19.25 Hz (double pole)
20.269 Hz (double pole)
22.8578 Hz (double pole)

You will have missed the corner frequencies below 0.1 Hz since they are not visible

in the frequency response function but will have started out with an ω^3 slope above 0.2 Hz. Also you will probably have approximated the strong decay around 20 Hs differently.

Chapter 4

Solution 4.1 The time derivative of (4.26) is

$$
\frac{dx_r(t)}{dt} = -\varepsilon x_{r0} e^{-\varepsilon t} \left(c_1 \sin\left(\sqrt{\omega_0^2 - \varepsilon^2}\, t\right) + c_2 \cos\left(\sqrt{\omega_0^2 - \varepsilon^2}\, t\right) \right)
$$
$$
+ x_{r0} e^{-\varepsilon t} c_1 \sqrt{\omega_0^2 - \varepsilon^2} \cos\left(\sqrt{\omega_0^2 - \varepsilon^2}\, t\right)
$$
$$
- x_{r0} e^{-\varepsilon t} c_2 \sqrt{\omega_0^2 - \varepsilon^2} \sin\left(\sqrt{\omega_0^2 - \varepsilon^2}\, t\right)
$$

With

$$
c_3 = -\left(c_1 \varepsilon + c_2 \sqrt{\omega_0^2 - \varepsilon^2} \right)
$$

$$
c_4 = -\left(c_2 \varepsilon + c_1 \sqrt{\omega_0^2 - \varepsilon^2} \right)
$$

we obtain

$$
\frac{dx_r(t)}{dt} = \varepsilon x_{r0} e^{-\varepsilon t} \left(c_3 \sin\left(\sqrt{\omega_0^2 - \varepsilon^2}\, t\right) + c_4 \cos\left(\sqrt{\omega_0^2 - \varepsilon^2}\, t\right) \right)
$$

As for the displacement seismometer, the amplitude ratio of two consecutive maxima or minima are solely determined by the exponential term. Therefore, equations (4.27) and (4.28) are valid for the determination of the damping constant.

Solution 4.2 a) The displacement impulse response is the inverse Fourier transform of:

$$T(j\omega) = \frac{Output_{disp}(j\omega)}{Input_{disp}(j\omega)}$$

By equivalence between differentiation in the time domain and multiplication with $j\omega$ in the frequency domain, the ground velocity response is related to the displacement frequency response by

$$Output_{vel}(j\omega) = j\omega \cdot Output_{disp}(j\omega) \quad \text{or}$$

$$Output_{disp}(j\omega) = \frac{1}{j\omega} \cdot Output_{vel}(j\omega)$$

Now let us examine the input signal, a step function in acceleration, which is equivalent to the integral of a spike. Hence, by equivalence between integration in the time domain and division by $j\omega$ in the frequency domain, the frequency response of the input signal in acceleration is:

$$Input_{acc}(j\omega) = \frac{1}{j\omega} \cdot 1$$

with 1 being the frequency response of an impulse (here in acceleration). In order to obtain the input signale in displacement, two times integration (division by $j\omega$) has to be performed:

$$Input_{disp}(j\omega) = \left(\frac{1}{j\omega}\right)^2 \cdot Input_{acc}(j\omega) = \left(\frac{1}{j\omega}\right)^3$$

Therefore,

$$T(j\omega) = \frac{Output_{disp}(j\omega)}{Input_{disp}(j\omega)} = \left(\frac{1}{j\omega} \cdot Output_{vel}(j\omega)\right) / \left(\frac{1}{j\omega}\right)^3$$

which becomes

$$T(j\omega) = (j\omega)^2 \cdot Output_{vel}(j\omega)$$

In the time domain, multiplication with $(j\omega)^2$ corresponds to double differentiation. Since high frequency noise will be greatly enhanced by differentiation, this explains why this approach to obtain the impulse response in practice often leads to difficulties.

b) From the peak amplitude values (0.0869349, -0.014175, 0.00231063, -0.000376618, 6.1422 x 10^{-5}) we obtain:

$$\left(\frac{a_1}{a_3} = 37.624\right) \Rightarrow \Lambda = 3.6276$$

$$\left(\frac{a_1}{a_2} = 6.13297\right) \Rightarrow \frac{\Lambda}{2} = 1.81368$$

From both values we obtain h = 0.5. From Fig. 4.4 we obtain for the period $T \approx 1.15$ sec (the exact value would be 1.15470054). This yields $f_0 = 1Hz$.

Solution 4.3 The pole positions for the three different damping factors are:

h = 0.25: -1.5708 ± 6.0837 i

h = 0.5: -3.14159 ± 5.4414 i

h = 0.62: -3.89557 ± 4.9298 i

The impulse responses and amplitude portion of the frequency responses (for a spike at point 100 out of 512, sampled at 20 Hz) are shown in look like Fig. A 4.1 and Fig. A 4.2, respectively.

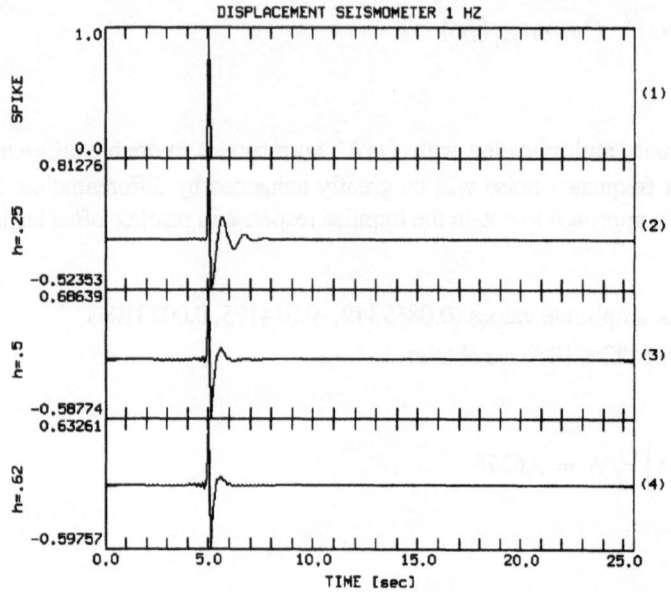

Fig. A 4.1 Impulse response functions for a displacement seismometer with three different damping factors. The impulsive input signal is shown in trace 1.

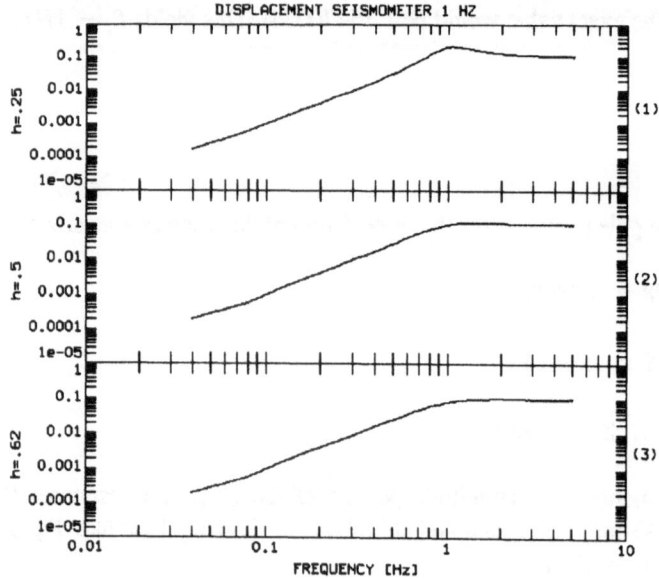

Fig. A 4.2 Amplitude portion of the frequency response functions for a displacement seismometer with 3 different damping factors.

Solution 4.4 The output signal has to be differentiated. This, however, corresponds to multiplication of the transfer function with s, a first degree polynomial with a single zero at the origin. Hence in order to simulate the output of an electrodynamic system, simply add a zero at the origin to the transfer function. For the damping factor h = 0.62, the resulting impulse response and amplitude frequency response are shown in Fig. A 4.3.

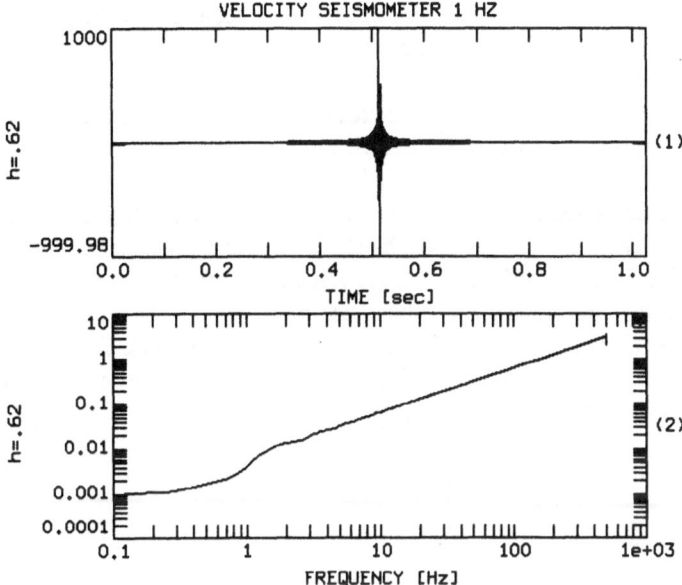

Fig. A 4.3 Impulse response function and corresponding amplitude frequency response function for a 1 Hz velocity seismometer. The reason for the apparent 'acausality' of the impulse response is the symmetry of the differentiation filter in the frequency domain, already discussed in the context of Solution 2.2.

Chapter 5

Solution 5.1 Using PITSA, we can attack this problem by simply checking out the effects of the discretization procedure for different signal frequencies of the 'continuous' signal. In order to create the traces in Fig. 5.1, the *test signal* tool from the *utilities* main menu option in PITSA has been used. They were created as noise-free sinusoids with a digitization frequency of 1024 Hz (to be high enough to appear continuous with respect to the discretization frequency of 10 Hz) and 2048 number of points per trace. For the first trace, the sinusoid parameters were 1 for the amplitude, 3.5 for the frequency and 0 for the phase (to be entered as: 1, 3.5, 0.0). The start time in this context is irrelevant. Hence the default value has been taken. You can use the same procedure to create input signals if arbitrary signal frequencies which subsequently can be discretized.

Using different input signal frequencies, you will notice that for a discretization frequency of 10 Hz, the reconstruction breaks down for frequencies around 5 Hz. For exactly 5 Hz, it strongly depends on the value of the phase angle of the continuous signal, if the discretization results in non-zero values. For a phase angle of exactly zero, the signal is discretized exactly at the zero crossings. Hence it cannot be distinguished from a zero signal. Therefore, for a unique reconstruction the input signals have to be below 1/2 of 10 Hz (= 5 Hz).

Solution 5.2 Imagine the frequency band as a band folded at 1/2 of the sampling frequency (folding frequency) The aliasing frequencies can then easily be obtained by vertical projection from a particular point on the frequency band down to the region between zero and folding frequency. Hence the aliasing frequency corresponding to 18.5 Hz and a sampling frequency of 10 Hz would be 1.5 Hz.

Frequency Band

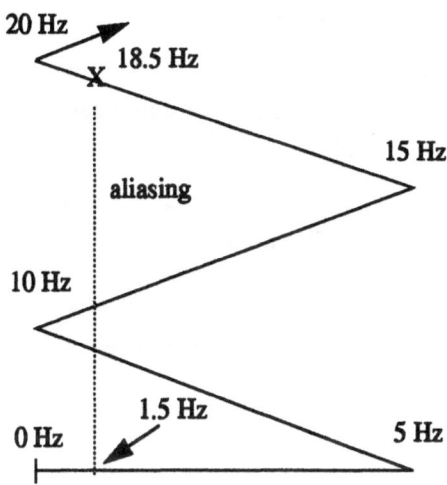

Fig. 5.1 Graphical illustration of the alias effect.

Chapter 6

The Wood-Anderson magnitude is defined as $M_{WA} = \log_{10}(A) - \log_{10}(A_0)$ with A being the amplitude measured on a Wood-Anderson displacement instrument and $-\log_{10}(A_0)$ being the distance correction which is exactly 3 for 100 km. The amplitude for a $M_{WA} = 0$ earthquake on this instrument will be $A_{M_{WA}=0} = 10^{0-3} = 0.001$. For the magnitude 6 earthquake this value will be $A_{M_{WA}=0} = 10^{6-3} = 1000$. Hence the dynamic range needed is at least $20 \cdot \log(10^6) = 120$ dB (> 20 bits).

Chapter 7

Solution 7.1 Since the inverse Fourier transform is evaluating the inverse z-transform on the unit circle, this means that the banded convergence region is chosen. Hence, it will correspond to a two-sided impulse response.

Solution 7.2 a) positive time shifts n_0 will cause a multiplication of the z-transform with a term z^{-n_0} which has a pole a at the origin $1/0^{n_0}$. b) negative time shifts n_0 will cause a multiplication of the z-transform with a term z^{n_0} which has a pole at infinity. Hence, depending on the sign of the shift, time shifts add poles at either the origin or at infinity of the z plane.

Solution 7.3 Let's assume we have two sequences with 1024 samples and a spike at position 1024 in each of them. If we convolve the two sequences in the time domain we expect the result to be a spike at position 2047. Hence, we have to pad up to this sample. If we have two sequences of different length, we have to pad both sequences up to one point less than twice the length of the longer sequence.

Chapter 8

Solution 8.1 In order to see how roots on the unit circle affect the properties of the impulse response of FIR filters, we first consider an impulse response function of length $M + 1 = 3$. Its z-transform corresponds to a simple quadratic

$$q + pz + z^2 \quad \text{with the two roots} \quad z_{1,2} = -\frac{p}{2} \pm \sqrt{\frac{p^2}{4} - q} = r e^{\pm i \Phi} \quad \text{for which}$$

the two following properties hold: $z_1 + z_2 = -p$ and $z_1 \cdot z_2 = q$.

For complex conjugate roots $r e^{\pm i \phi}$ it follows that $q = r^2$. Using Euler's formulas, we obtain that $2r \cos \Phi = -p$. Hence, the discrete finite impulse response for a system with the two complex conjugate roots $r e^{\pm i \phi}$ will consist of the triplet

$$(r^2, -2r \cos \phi, 1)$$

If the roots are on the unit circle ($r = 1$), this will result in a symmetric impulse response. This will also be the case for higher order systems which only contain zeros on the unit circle since the convolution of symmetric impulse response functions stays symmetric. Because of this symmetry, roots on the unit circle have to be corrected for, if the goal is to remove all noncausal filter effects from digital seismic recordings. What makes it even more complicated is the fact, that a system with only roots on the unit circle has only a single waveform representation for the given amplitude spectrum. This again can be seen from the example of the simple quadratic. The square of the z-transform amplitude spectrum for the last equation equals

$$(r^2 - 2r\cos\phi z + z^2) \cdot (z^2 - 2r\cos\phi z + r^2)$$

with the four roots $z_{1,2} = re^{\pm i\Phi}$ and $z_{3,4} = 1/r \cdot e^{\mp i\phi}$. For $r = 1$ we end

up with two identical pairs of complex conjugate roots $z_{1,2} = z_{3,4} = e^{\pm i\Phi}$.

Since any real sequence corresponding to the z-transform amplitude spectrum has to be represented by a pair of complex conjugate roots, we see immediately that only one waveform representation exists for systems having only roots on the unit circle. Therefore, we cannot split up these system components in FIR filters exactly into minimum and maximum phase partial systems.

If we assume these zeros belonging to the minimum phase part of the filter as it can be sometimes found in textbooks, this would be equivalent to simply ignoring them for the purpose of removing the acausal filter response. Depending on the number of zeros on the unit circle involved, this would cause ignoring an important contribution to the acausality. Consequently the correction filter would perform very poorly. On the other hand, if we treat them as belonging to the maximum phase part, they will cause an instability during the inversion because the roots will become poles in the correcting filter. One way out of this dilemma is to approximate such a FIR impulse response function in terms of "equivalent" minimum and maximum phase parts without involving zeros on the unit circle (Scherbaum et al., 1994).

Appendix B: Listing of Source Codes

Warranty Disclaimer: The following source code listings are included for educational purposes. They are provided without warranty of any kind, either expressed or implied, including but not limited to the fitness for a particular purpose.

Listing 1 mkcausal.c

```
/**
  NAME: mkcausal
  SYNOPSIS:
    mkcausal <flags>
    flags: -w <filter wavelet file>
           -i <data input file>
           -o <output file>
           -z <no. of zeros to append to input file. Optional!>
           -c Corrects for linear phase of input wavelet. Optional!
  VERSION: 1.0
  DATE: 1993-04-12 (Frank Scherbaum)
  DESCRIPTION: From the coefficients of a FIR filter in the <filter wavelet file> a recursive cor-
  rection filter is constructed to remove the acausal part of the impulse response from seismic
  recording which have been filtered with the FIR filter. The input data are assumed to be in the
  <data input file>. The corrected trace  is written into the <output file> All files are 1 value per
  row ASCII text files. Using a -z <n> option, <n> number of zeroes are appended to the data
  from the  input file on input. This option prevents the supression of cutting off data from the
  output file in case the response of the correction filter is longer than the  input trace (e.g. if a
  FIR filter response itself is being corrected).

**/
/* begin Makefile defines   */
/* define DBL           for double  precision calculation */
/* define SNG           for single precision calculation */
/* define MASSAGE_CIRCLE    for shifting roots from circle */
/*                  this is necessary for double precision calculation */
/* end  Makefile defines    */
```

```c
#ifndef DBL
#    define SNG              /* default precision */
#endif


#define NMAX 100                 /* maximum wavelet length for rooting */
#define ONE Complex(1.0,0.0)
#define PI 3.141592653589793
#define SHIFT_ROOT 1.000001      /* factor by which unit circle roots are shifted */

#include <stdio.h>
#include <stdlib.h>
#include <stddef.h>
#include <math.h>
#include <ctype.h>
#include <string.h>
#include <malloc.h>
#include <memory.h>
#include "nrutil.h"
#include "complex.h"
main(argc,argv)
    int argc;
    char   *argv[];

{
    FILE *fo;               /* FILE pointer output file        */
    FILE *fi;               /* FILE pointer input file         */
    char fir_name[80];         /* input file name FIR filter      */
    char in_name[80];          /* input file name trace file      */
    char out_name[80];         /* output file name                */
    int i,j,k;              /* indeces                   */
    int nskip = 0;          /* no. of lines to skip on input   */
    char ch;                /* char. buffer                */
    int ntot;               /* total no. of lines in file    */
#ifdef SNG
    float *wv;              /* wavelet                     */
    float x_inp;            /* input float buffer              */
    float re[NMAX],im[NMAX];   /* real and imaginary parts of roots[] */
    float shift_root;       /* multipl. factor for shifting roots */
                            /* away from unit circle       */
#endif
#ifdef DBL
    double *wv;             /* wavelet                     */
    double x_inp;           /* input float buffer              */
    double re[NMAX],im[NMAX];  /* real and imaginary parts of roots[] */
    double shift_root;      /* multipl. factor for shifting roots */
#endif
    int ndat;               /* length of wavelet           */
    int wv_m;               /* degree of wavelet polynomial    */
    char text_buffer[255];     /* input text buffer             */
    char *string_end;          /* pointer to possible CR in buffer   */
    float freq;             /* frequency in Hz             */
    fcomplex roots[NMAX];      /* roots of input wavelet          */
    float *xf;              /* DFT (Numerical Recipes format)   */
    int nfft;               /* no. of points for FFT           */
```

```
   float fdig;              /* digitization frequency in Hz        */
   fcomplex z;                 /* complex dummy variables for inverse */
   fcomplex dum_1;             /* complex dummy variable  for inverse */
                          /* z transform calculation         */
   float max;                  /* scaling factor for polynomial       */
   fcomplex dft_spec;          /* z-transform values on unitcircle    */
   float *min_phas, *max_phas; /* minimum phase and maximum phase     */
                          /* components of wavelet            */
   int m_min,m_max;            /* degree of minimum and maximum phase */
                          /* portions of wavelet polynomial    */
   float *equ_phas;            /* equi delay part of wavelet        */
   int m_equ;                  /* degree of equi delay part        */
   int pad_zeros;              /* no. of padded zeroes for filtering  */
   int lead_zeros;             /* no. of leading zeroes automatically */
                          /* put before input trace           */
   int n,m;                    /* degrees of ARMA coeeficienst      */
   int ndat2;                  /* length of output file           */
   float max_dist;             /* maximum root distance from origin   */
#ifdef SNG
   float *x2;               /* input trace                  */
   float *x;                /* time reversed input trace         */
   float *y;                /* filtered output trace          */
   float *a,*b;             /* ARMA coefficients              */
#endif
#ifdef DBL
   double *x2;                 /* input trace                  */
   double *x;                  /* time reversed input trace       */
   double *y;                  /* filtered output trace          */
   double *a,*b;               /* ARMA coefficients              */
#endif
   float *temp;                /* buffer for time shift           */
   int max_pos;                /* index of maximum point of FIR filter*/
   float shift_samples;        /* time shift in samples caused by the */
                          /* causal linear phase FIR filter:    */
                          /* = (float)(ndat-1)/2.0          */
   int correct_time;           /* if 1 -> time correction for linear  */
                          /* phase performed, else not        */
   int dont_shift0;            /* if 1 -> zeroes on UC are not shifted*/
   float len;                  /* root distance from origin        */
   pad_zeros = 0;
   correct_time = 0;
   dont_shift0 = 0;
   for (i=1; i<argc;i++)
   {
      if (argv[i][0] == '-' && strlen(argv[i]) >= 2 )
      {
         switch( argv[i][1])
         {
            case 'w': /* FIR filter input file name */
            case 'W': /* FIR filter input file name */
            {
               i++;
               strcpy(fir_name,&argv[i][0]);
```

```
                break;
            }
            case 'i': /* input file name */
            case 'I': /* input file name */
            {
                i++;
                strcpy(in_name,&argv[i][0]);
                break;
            }
            case 'o': /* output file name */
            case 'O': /* output file name */
            {
                i++;
                strcpy(out_name,&argv[i][0]);
                break;
            }
            case 'z': /* no. of zeros to append on input of data file */
            case 'Z':
            {
                i++;
                pad_zeros = atoi(&argv[i][0]);
                break;
            }
            case 'c': /* correct for linear phase shift */
            case 'C':
            {
                correct_time = 1;
                break;
            }
            case 'n': /* correct for linear phase shift */
            case 'N':
            {
                dont_shift0 = 1;
                break;
            }
        }
    }
}

if ((argc < 3) ||
    (strlen(fir_name) == 0)||
    (strlen(in_name) == 0)||
    (strlen(out_name) == 0))
{
    printf("USAGE: mkcausal <flags>\n");
    printf("flags: -w <filter wavelet file>\n");
    printf("       -i <data input file>\n");
    printf("       -o <output file>\n");
    printf("       -z <no. of zeros to append to input file. Optional!>\n");
    printf("       -c Corrects for linear phase of input wavelet. Optional!\n");
    exit(1);
}
/* estimate no. of lines in file */
```

```
  ntot  = 0;
  if ((fo = fopen(fir_name,"rt")) != NULL)
  {
       while((ch=fgetc(fo)) != EOF)
       {
          if (ch == '\n')
             ntot++;
       }
     fclose(fo);
  }
  ndat = ntot-nskip;
  shift_samples = (float)(ndat -1)/2.;  /* linear phase shift caused */
                            /* by this filter        */
  /* wavelet polynomial */
#ifdef SNG
  wv  = (float *)calloc(ndat,sizeof(float));
#endif
#ifdef DBL
  wv  = (double *)calloc(ndat,sizeof(float));
#endif
  /* INPUT  wavelet */
  if((fi = fopen(fir_name,"rt")) == NULL)
  {
       printf("input file %s cannot be opened!\n",in_name);
       exit(1);
  }

  for(k=0;k<ndat+nskip;k++)
  {
     if(fgets(text_buffer,255,fi) != NULL)
     {
        if ((string_end = memchr(&text_buffer[0],'\n',255)) != NULL)
             *string_end = '\0';
#ifdef SNG
        sscanf(text_buffer,"%f",&x_inp);
#endif
#ifdef DBL
        sscanf(text_buffer,"%lf",&x_inp);
#endif
     }
     else
       x_inp = 0.0;
     if(k >= nskip)
       *(wv+k-nskip) = x_inp;
  }
  fclose(fi);
  /* scale trace so that the coefficient to     */
  /* the highest power of z (last) becomes 1     */
  wv_m = ndat-1;
  /* find position of maximum of FIR filter      */
  max_pos = 0;
  max =wv[0];
  for(k=0;k<ndat;k++)
```

```
    {
      if(wv[k] > max)
      {
        max = wv[k];
        max_pos = k;
      }
    }
    /* scale polynomial so that coefficient to largest */
    /* power of z becomes 1                            */
    max = wv[wv_m];
    for(k=0;k<ndat;k++)
      wv[k] /= max;
    printf("coefficient[%d]: %f\n",wv_m,max);
    /* get roots in z                 */
#ifdef SNG
    zrhqr(wv,wv_m,re,im);
#endif
#ifdef DBL
    zrhqrdb(wv,(long)wv_m,re,im);
#endif
    free((char *)wv);
    /* output                         */
    if((fo = fopen("roots","wt")) == NULL)
    {
        printf("output file for roots cannot be opened!\n");
        exit(1);
    }
    max_dist = 0.0;
    /* complex roots of polynomial in z           */
    for(i=1;i<=wv_m;i++)
    {
#ifdef SNG
      roots[i].r = re[i];
      roots[i].i = im[i];
#endif
#ifdef DBL
      roots[i].r = (float)re[i];
      roots[i].i = (float)im[i];
#endif

      fprintf(fo,"% 12f % 12f\n",roots[i].r,roots[i].i);
      if (Cabs(roots[i]) > max_dist)
        max_dist = Cabs(roots[i]);
    }
    fclose(fo);
    printf("Most distant root at distance %f\n",max_dist);
    /* Inverse z transform */
    /* calculate next power of 2 */
    nfft = 1;
    while(nfft < ndat)
      nfft*=2;
    fdig = 1;
    /* determine equi delay part */
```

```
equ_phas = vector(1L,(long)nfft);
for(j=1;j<=nfft/2+1;j++)
{
   freq  = (j-1) * fdig / nfft;
   /* construct DFT from roots of wavelet */
   dft_spec = ONE;
   m_equ=0;
   if(j==1)
   {
      /* output                     */
      if((fo = fopen("roots.equ","wt")) == NULL)
      {
           printf("output file roots.equ cannot be opened!\n");
           exit(1);
      }
   }
   for(i=1;i<=wv_m;i++)
   {
      z=Complex(cos(2.*PI*freq),sin(2.*PI*freq));
      if (Cabs(roots[i]) == 1.0)
      {
         if (j == 1){
            fprintf(fo,"% 12f  % 12f\n",roots[i].r,roots[i].i);
         }
         m_equ++;
         dum_1 = Csub(z,roots[i]);
         dft_spec=Cmul(dft_spec,dum_1);
      }
   }
   if (j == 1)
      fclose(fo);

   /* fill spectrum */
   if (j == 1) /* frequency 0.0 */
   {
      equ_phas[1]= dft_spec.r;
   }
   else if (j == nfft/2+1) /* Nyquist frequency 0.0 */
   {
      equ_phas[2]= dft_spec.r;

   }
   else
   {
      equ_phas [2*j-1] =  dft_spec.r;
      equ_phas [2*j]   =  dft_spec.i;
   }
}
/* inverse  FFT                   */
realft(equ_phas,(long)nfft,-1);
/* scale back to  original amplitudes */
for(j=1;j<=nfft;j++)
{
```

```c
        equ_phas[j] /= (float)(nfft/2);
  }
  printf("Wavelet has %d roots on unit circle.\n",m_equ);
  /* output                              */
  if((fo = fopen("wavelet.equ","wt")) == NULL)
  {
        printf("output file wavelet.equ cannot be opened!\n");
        exit(1);
  }
  for(j=1;j <= m_equ+1;j++)
  {
        fprintf(fo,"%g\n",equ_phas[j]);

  }
  fclose(fo);
#ifdef MASSAGE_CIRCLE
  if(dont_shift0 != 1)
  {
    shift_root = 1./SHIFT_ROOT; /* shift first root to max. phase part */
    /* massage the roots on the unit circle to either inside or outside */
    for(i=1;i<=wv_m;i++)
    {
       if (Cabs(roots[i]) == 1.0)
       {
         /* check if it is a real or a complex root */
         if ( (i < wv_m) &&
            (roots[i].r == roots[i+1].r) &&
            (roots[i].i == -roots[i+1].i)
            ){ /* complex root */
         /* shift the root opposite from the last on */
         roots[i].r *= shift_root; roots[i].i *= shift_root;
         roots[i+1].r *= shift_root; roots[i+1].i *= shift_root;
         if(shift_root == SHIFT_ROOT) shift_root = 1./SHIFT_ROOT;
         else
            shift_root = SHIFT_ROOT;

         } else { /* real root */
            roots[i].r *= shift_root; roots[i].i *= shift_root;
            if(shift_root == SHIFT_ROOT) shift_root = 1./SHIFT_ROOT;
            else
              shift_root = SHIFT_ROOT;
         }
       }
    }
    printf("%d roots shifted from UC by a factor of %f\n",m_equ,SHIFT_ROOT);
  }
#endif
  /* determine maximum phase part */
  max_phas = vector(1L,(long)nfft);
  for(j=1;j<=nfft/2+1;j++)
  {
    freq  = (j-1) * fdig / nfft;
    /* construct DFT from roots of wavelet */
```

```
dft_spec = ONE;
m_max=0;
if(j==1)
{
   /* output                          */
   if((fo = fopen("roots.max","wt")) == NULL)
   {
        printf("output file roots.max cannot be opened!\n");
        exit(1);
   }
}
for(i=1;i<=wv_m;i++)
{
   z=Complex(cos(2.*PI*freq),sin(2.*PI*freq));
   if (Cabs(roots[i]) < 1.0)
   {
      if (j == 1)
         fprintf(fo,"% 12f % 12f\n",roots[i].r,roots[i].i);
      m_max++;
      dum_1 = Csub(z,roots[i]);
      dft_spec=Cmul(dft_spec,dum_1);
   }
}
if (j == 1)
   fclose(fo);

/* fill spectrum */
if (j == 1) /* frequency 0.0 */
{
   max_phas[1]= dft_spec.r;
}
else if (j == nfft/2+1) /* Nyquist frequency 0.0 */
{
   max_phas[2]= dft_spec.r;

}
else
{
   max_phas [2*j-1] =  dft_spec.r;
   max_phas [2*j]   =  dft_spec.i;
}
}
/* inverse  FFT              */
realft(max_phas,(long)nfft,-1);
/* scale back to  original amplitudes */
for(j=1;j<=nfft;j++)
{
     max_phas[j]  /= (float)(nfft/2);
}
printf("Wavelet has %d roots in maximum phase part.\n",m_max);
/* output                   */
if((fo = fopen("wavelet.max","wt")) == NULL)
{
```

```c
            printf("output file wavelet.max cannot be opened!\n");
            exit(1);
   }
   for(j=1;j<=m_max+1;j++)
   {

            fprintf(fo,"%g\n",max_phas[j]);


   }
   fclose(fo);
   /* determine minimum phase part */
   min_phas = vector(1L,(long)nfft);
   for(j=1;j<=nfft/2+1;j++)
   {
      freq  = (j-1) * fdig / nfft;
      /* construct DFT from roots of wavelet */
      dft_spec = ONE;
      m_min=0;
      if(j==1)
      {
         /* output                    */
         if((fo = fopen("roots.min","wt")) == NULL)
         {
               printf("output file roots.min cannot be opened!\n");
               exit(1);
         }
      }
      for(i=1;i<=wv_m;i++)
      {
         z=Complex(cos(2.*PI*freq),sin(2.*PI*freq));
         if (Cabs(roots[i]) > 1.0)
         {
            if (j == 1)
               fprintf(fo,"% 12f % 12f\n",roots[i].r,roots[i].i);
            m_min++;
            dum_1 = Csub(z,roots[i]);
            dft_spec=Cmul(dft_spec,dum_1);
         }
      }
      if (j == 1)
         fclose(fo);

      /* fill spectrum */
      if (j == 1) /* frequency 0.0 */
      {
         min_phas[1]= dft_spec.r;
      }
      else if (j == nfft/2+1) /* Nyquist frequency 0.0 */
      {
         min_phas[2]= dft_spec.r;

      }
      else
      {
```

```c
            min_phas [2*j-1] =  dft_spec.r;
            min_phas [2*j]   =  dft_spec.i;
        }
    }
    /* inverse  FFT                    */
    realft(min_phas,(long)nfft,-1);
    /* scale back to  original amplitudes */
    for(j=1;j<=nfft;j++)
    {
        min_phas[j]  /= (float)(nfft/2);
    }
    printf("Wavelet has %d roots in minimum phase part.\n",m_min);
    /* output                    */
    if((fo = fopen("wavelet.min","wt")) == NULL)
    {
        printf("output file wavelet.min cannot be opened!\n");
        exit(1);
    }
    for(j=1;j <= m_min+1;j++)
    {
        fprintf(fo,"%g\n",min_phas[j]);

    }
    fclose(fo);
/*
    GENERAL DIFFERENCE EQUATION:
```

$$
y[i] = \sum_{k=1}^{m} a*y[i-k] + \sum_{k=0}^{n} b*x[i-k]
$$

```
     \----------/   \------------/
         AR              MA
```

Since there is no a term, the first a term is treated as a
            0   1

If the moving average parts should be ignored, a single 0 should
be given as only MA coefficients.

y[] = output trace
x[] = trace to be filtered

```c
*/
    /* ARMA coefficients                */
#ifdef SNG
    b = (float *)calloc(m_max+1,sizeof(float));
```

```c
       a = (float *)calloc(m_max+1,sizeof(float));
#endif
#ifdef DBL
   b = (double *)calloc(m_max+1,sizeof(double));
   a = (double *)calloc(m_max+1,sizeof(double));
#endif
   for(j=1;j<=m_max+1;j++)
   {
#ifdef SNG
         b[j-1] = max_phas[j];
#endif
#ifdef DBL
         b[j-1] = (double)max_phas[j];
#endif
   }
   /* Now, the time reversed maximum phase part of the */
   /* FIR filter is on b[0],.....,b[m_max]          */
   /* INPUT: trace to be filtered   */
   /* estimate no. of lines in file */
   ntot  = 0;
   if ((fo = fopen(in_name,"rt")) != NULL)
   {
         while((ch=fgetc(fo)) != EOF)
         {
            if (ch == '\n')
               ntot++;
         }
      fclose(fo);
   }
   lead_zeros = 2*wv_m;
   ndat2 = ntot - nskip + lead_zeros + pad_zeros;
#ifdef SNG
   x  = (float *)calloc(ndat2,sizeof(float));
   x2 = (float *)calloc(ndat2,sizeof(float));
#endif
#ifdef DBL
   x  = (double *)calloc(ndat2,sizeof(double));
   x2 = (double *)calloc(ndat2,sizeof(double));
#endif
   /* INPUT data file */
   if((fi = fopen(in_name,"rt")) == NULL)
   {
         printf("input file %s cannot be opened!\n",in_name);
         exit(1);
   }
   for(k=0;k<ndat2-nskip-pad_zeros-lead_zeros;k++)
   {
      if(fgets(text_buffer,255,fi) != NULL)
      {
         if ((string_end = memchr(&text_buffer[0],'\n',255)) != NULL)
             *string_end = '\0';
#ifdef SNG
         sscanf(text_buffer,"%f",&x_inp);
```

```
#endif
#ifdef DBL
        sscanf(text_buffer,"%lf",&x_inp);
#endif
    }
    else
      x_inp = 0.0;

    if(k >= nskip)
      x2[k-nskip+lead_zeros] = x_inp;
  }
  fclose(fi);
  /* flip in time */
  for(k=0;k<ndat2;k++)
  {
    x[k]=x2[ndat2-1-k];
  }
  /* output trace                    */
#ifdef SNG
  y = (float *)calloc(ndat2,sizeof(float));
#endif
#ifdef DBL
  y = (double *)calloc(ndat2,sizeof(double));
#endif
  m = m_max;
  n = m_max;
  for (i=0; i<=n; i++)
    b[i] /= b[n];
  for (i=1; i<=m; i++)
    a[i-1] = -b[m-i]/b[m];
  /* filter                      */
  for (i=0; i<ndat2; i++) {
    y[i] = 0.0;
    /* MA */
    for (k=0; k<=n; k++) {
      if ((i-k) >= 0) {
        y[i] += x[i-k]*b[k];
      }
    }
    /* AR */
    for (k=1; k<=m; k++) {
      if ((i-k) >= 0) {
        y[i] += y[i-k]*a[k-1];
      }
    }
  }
  temp = (float *)calloc(ndat2,sizeof(float));
  /* flip back in time */
  for(k=0;k<ndat2;k++)
  {
    temp[k]=(float)y[ndat2-1-k];
  }
  if (correct_time == 1)
```

```
    {
        time_shift(temp,ndat2,shift_samples);
        printf("linear phase correcction: %f samples\n",shift_samples);
    }
    else
        printf("Trace needs linear phase correcction: %f samples\n",shift_samples);
    /* output                        */
    if((fo = fopen(out_name,"wt")) == NULL)
    {
        printf("output file %s cannot be opened!\n",out_name);
        exit(1);
    }
    for(j=0;j<ndat2-lead_zeros;j++)
    {
        fprintf(fo,"%g\n",temp[j+lead_zeros]);

    }
    fclose(fo);
    /* reconstruct original wavelet */
    xf  = vector(1L,(long)nfft);
    for(j=1;j<=nfft/2+1;j++)
    {
        freq  = (j-1) * fdig / nfft;
        /* construct DFT from roots of wavelet */
        dft_spec = ONE;
        m_max=0;
        for(i=1;i<=wv_m;i++)
        {
            z=Complex(cos(2.*PI*freq),sin(2.*PI*freq));
            {
                m_max++;
                dum_1 = Csub(z,roots[i]);
                dft_spec=Cmul(dft_spec,dum_1);
            }
        }
        /* fill spectrum */
        if (j == 1) /* frequency 0.0 */
        {
            xf[1]= dft_spec.r;
        }
        else if (j == nfft/2+1) /* Nyquist frequency 0.0 */
        {
            xf[2]= dft_spec.r;

        }
        else
        {
            xf [2*j-1] =  dft_spec.r;
            xf [2*j]  =  dft_spec.i;
        }
    }
    /* inverse  FFT                    */
    realft(xf,(long)nfft,-1);
```

```
    /* scale back to  original amplitudes */
    for(j=1;j<=nfft;j++)
    {
          xf[j]  /= (float)(nfft/2);
          xf[j]  *= max;

    }
    /* output                         */
    if((fo = fopen("wavelet.rec","wt")) == NULL)
    {
          printf("output file wavelet.rec cannot be opened!\n");
          exit(1);
    }
    for(j=1;j<=wv_m+1;j++)
    {
          fprintf(fo,"%g\n",xf[j]);

    }
    fclose(fo);
    /* free allocated memory              */
    free_vector(max_phas,1L,(long)nfft);
    free_vector(min_phas,1L,(long)nfft);
    free_vector(equ_phas,1L,(long)nfft);
    free_vector(xf,1L,(long)nfft);
    free((char *)a);
    free((char *)b);
    free((char *)x);
    free((char *)x2);
    free((char *)y);
    free((char *)temp);

}
/**
   NAME: time_shift
   SYNOPSIS:
   float *y;
   int ndat;
   float shift_samples;
   time_shift(y,ndat,extra_samples);
   DESCRIPTION: Performs a time shift in the frequency domain by
   multiplication with the corresponding phase shift operator.
   DATE: July 2, 1993 (Frank Scherbaum)
**/
int time_shift(y,ndat,shift_samples)
float *y;
int ndat;
float shift_samples;
{
   float *shift;         /* spectrum corresponding to the desired time shift */
   int i,j;
   float amp,phase;       /* amplitude and phase of shifting spectrum */
   float real, imag;     /* real, imaginary of shifting spectrum */
   float t_samp;          /* sampling interval */
```

```
int ia;                /* integer part of shift */
float extra_samples;   /* fraction of samples to shift */
float *b1;             /* trace buffers */
float *bb1, *bb2;      /* trace buffers */
float x1;              /* dummy variable */
int nfft;
/* allocate buffer */
b1 = (float *)calloc(ndat,sizeof(float));
ia = 0;
if(shift_samples == (int)shift_samples)
{ /*integer multiple of 1 sample */
   ia = (int)shift_samples;
   extra_samples = 0.0;
}
else
{
   ia = (int)shift_samples;
   extra_samples = shift_samples - ia;
}
if(ia >0)
{
   for (j = ndat-1 ; j >= ia; j--)
   {
      x1 = *(y + j - ia);
      *(b1+j) = x1;
   }
   for (j = 0; j < ia; j++)
      *(b1+j) = 0.0;
}
else if(ia < 0)
{
   for (j = 0 ; j < ndat +ia -1 ; j++)
   {
      x1 = *(y + j - ia);
      *(b1+j) = x1;
   }
   for (j = ndat-ia ; j <ndat; j++)
   {
      *(b1+j) = 0.0;
   }
}
else if (ia == 0)
{
   for (j = 0; j < ndat; j++)
      *(b1+j) = *(y + j);
}
/*
   every shift less than a sample is done in the
   frequency domain
*/
if (extra_samples != 0.0) /* non-integer part of shift */
{
   nfft = 1;
```

```
    while(nfft < ndat)
        nfft *= 2;
    nfft *= 2; /* to avoid wrap around */
    bb1 = (float *)calloc(nfft,sizeof(float));
    bb2 = (float *)calloc(nfft,sizeof(float));
    for(j=0;j<ndat;j++)
    {
        bb1[j] = b1[j];
    }
    realft(bb1-1,(long)nfft,1);
    amp = 1.0;
    /*
    1 sample shift ==
    phase of PI at f Nyquist (linear in between)
    */
    bb2[0] = amp;
    bb2[1] = amp;
    for (j=1; j < nfft/2; j++)
    {
        phase = extra_samples*
            ((float)j/(float)nfft)*PI;
        real = amp*cos(phase);
        imag = amp*sin(phase);
        bb2[2*j] = real;
        bb2[2*j+1] = imag;
    }
    /* spectral multiplication */
    bb1[0] *= bb2[0];
    bb1[1] *= bb2[1];
    for (j=1; j < nfft/2; j++)
    {
        real = bb1[2*j]*bb2[2*j] - bb1[2*j+1]*bb2[2*j+1];
        imag = bb1[2*j]*bb2[2*j+1] + bb1[2*j+1]*bb2[2*j];
        bb1[2*j] = real;
        bb1[2*j+1] = imag;
    }
    /* inverse FFT */
    realft(bb1-1,(long)nfft,-1);
    /* scale amplitudes back */
    for(j=0;j<ndat;j++)
    {
        y[j] = 2*bb1[j]/nfft;
    }
    free((char *)b1);
    free((char *)bb1);
    free((char *)bb2);
    }
}
```

Listing 2 Makefile for mkcausal

```
DEF = -DDBL -DMASSAGE_CIRCLE
CFLAGS = -g $(DEF)
.c.o:
cc -c $(CFLAGS) $*.c

mkcausal:   mkcausal.o zrhqr.o balanc.o nrutil.o hqr.o complex.o realft.o four1.o
cc $(CFLAGS) -o mkcausal mkcausal.o zrhqr.o balanc.o nrutil.o hqr.o complex.o realft.o
four1.o -lm
```

All routines other than mkcausal.c that are needed in this context are from Press et al. (1992).

Index

158

Springer-Verlag
and the Environment

We at Springer-Verlag firmly believe that an international science publisher has a special obligation to the environment, and our corporate policies consistently reflect this conviction.

We also expect our business partners – paper mills, printers, packaging manufacturers, etc. – to commit themselves to using environmentally friendly materials and production processes.

The paper in this book is made from low- or no-chlorine pulp and is acid free, in conformance with international standards for paper permanency.

Lecture Notes in Earth Sciences

Vol. 37: A. Armanini, G. Di Silvio (Eds.), Fluvial Hydraulics of Mountain Regions. X, 468 pages. 1991.

Vol. 38: W. Smykatz-Kloss, S. St. J. Warne, Thermal Analysis in the Geosciences. XII, 379 pages. 1991.

Vol. 39: S.-E. Hjelt, Pragmatic Inversion of Geophysical Data. IX, 262 pages. 1992.

Vol. 40: S. W. Petters, Regional Geology of Africa. XXIII, 722 pages. 1991.

Vol. 41: R. Pflug, J. W. Harbaugh (Eds.), Computer Graphics in Geology. XVII, 298 pages. 1992.

Vol. 42: A. Cendrero, G. Lüttig, F. Chr. Wolff (Eds.), Planning the Use of the Earth's Surface. IX, 556 pages. 1992.

Vol. 43: N. Clauer, S. Chaudhuri (Eds.), Isotopic Signatures and Sedimentary Records. VIII, 529 pages. 1992.

Vol. 44: D. A. Edwards, Turbidity Currents: Dynamics, Deposits and Reversals. XIII, 175 pages. 1993.

Vol. 45: A. G. Herrmann, B. Knipping, Waste Disposal and Evaporites. XII, 193 pages. 1993.

Vol. 46: G. Galli, Temporal and Spatial Patterns in Carbonate Platforms. IX, 325 pages. 1993.

Vol. 47: R. L. Littke, Deposition, Diagenesis and Weathering of Organic Matter-Rich Sediments. IX, 216 pages. 1993.

Vol. 48: B. R. Roberts, Water Management in Desert Environments. XVII, 337 pages. 1993.

Vol. 49: J. F. W. Negendank, B. Zolitschka (Eds.), Paleolimnology of European Maar Lakes. IX, 513 pages. 1993.

Vol. 50: R. Rummel, F. Sansò (Eds.), Satellite Altimetry in Geodesy and Oceanography. XII, 479 pages. 1993.

Vol. 51: W. Ricken, Sedimentation as a Three-Component System. XII, 211 pages. 1993.

Vol. 52: P. Ergenzinger, K.-H. Schmidt (Eds.), Dynamics and Geomorphology of Mountain Rivers. VIII, 326 pages. 1994.

Vol. 53: F. Scherbaum, Basic Concepts in Digital Signal Processing for Seismologists. X, 158 pages. 1994.